Kai Sandfort

The Factorization Method for Inverse Scattering from Periodic Inhomogeneous Media

The Factorization Method for Inverse Scattering from Periodic Inhomogeneous Media

by
Kai Sandfort

KIT Scientific Publishing

Dissertation, Karlsruher Institut für Technologie
Fakultät für Mathematik,
Tag der mündlichen Prüfung: 03.02.2010

Impressum

Karlsruher Institut für Technologie (KIT)
KIT Scientific Publishing
Straße am Forum 2
D-76131 Karlsruhe
www.ksp.kit.edu

KIT – Universität des Landes Baden-Württemberg und nationales
Forschungszentrum in der Helmholtz-Gemeinschaft

KIT Scientific Publishing 2010
Print on Demand

ISBN 978-3-86644-550-5

The Factorization Method for Inverse Scattering from Periodic Inhomogeneous Media

Zur Erlangung des akademischen Grades eines

DOKTORS DER NATURWISSENSCHAFTEN

von der Fakultät für Mathematik des
Karlsruher Instituts für Technologie (KIT)
genehmigte

DISSERTATION

von

Dipl.-Math. Kai Sandfort

aus Gummersbach

Tag der mündlichen Prüfung: 03. Februar 2010
Referent: Prof. Dr. Andreas Kirsch
Korreferent: PD Dr. Frank Hettlich

Preface

In this work, we consider the scattering of acoustic and electromagnetic waves from penetrable periodic structures in $\mathbb{R}^3$. The periodicity is manifested here in the physical parameter of the scatterer which controls the propagation of the incident field and is given as a function of the three spatial coordinates. The emphasis of our research is on the geometric reconstruction of the scatterer by means of scattering data collected in its near field region. Since this task is 'inverse' to the 'direct scattering problem' of computing the scattered field caused by a specified structure, it is called a problem of 'inverse scattering'. Both of these problems are of high and rapidly growing practical relevance. The focus is certainly set on the case of an electromagnetic incidence. A major driving force for this is made up by the technical advances in the fabrication of optical devices in the last thirty years, which opened a broad range of exciting new applications and functions. Processes from the semiconductor industry allow to produce structural features on the lengthscale of electromagnetic waves, in particular light ($\approx$ 380 nm$-$780 nm). In this so-called 'resonance region' of similar lengthscales, the radiation interacts with the structure in a complicated way which prohibits an accurate representation by geometrical optics. In fact, it necessitates to deal with the full electromagnetic vector-field equations, the Maxwell's equations. Tackling this challenge makes it possible nowadays e.g. to design diffraction elements, frequency filters, and waveguides. At the same time, there is an increasing demand for modeling and simulating the propagation of acoustic as well as electromagnetic radiation in complex microscopic and macroscopic settings. On each lengthscale, there are evident and pressing problems of 'inverse scattering', referring to the reconstruction of certain features of a target structure from the knowledge of the generated scattered field. Applications which lead to such problems are found e.g. in medical imaging, near field microscopy, surface inspection, object detection and control, exploration geophysics, remote sensing, as well as in the iterative design of sonic and optical devices. It is the latter application which this thesis contributes to. We consider the important class of penetrable periodic devices, irradiated by either acoustic or electromagnetic fields. The main part of the thesis is devoted to the proof that the so-called *Factorization Method*, a well-known reconstruction technique, can be adapted to these settings in order to solve the corresponding inverse scattering problems. The central intention here is to let the governing material parameter be essentially unrestricted, taking into account the fast progress in material science and assembling techniques. In addition, we formulate an efficient numerical solver for the direct problem for smooth material parameters and devise a related, but new solver for piecewise constant parameters, as they characterize many of today's configurations. We combine these solvers with an implementation of the Factorization Method and validate our theoretical results in a couple of simulated scattering problems.

The work on this thesis has been supported by the German Research Foundation (DFG) through a grant within the program of the Research Training Group (GRK) 1294 "Analysis, Simulation and Design of Nanotechnological Processes" at the Department of Mathematics, Universität Karlsruhe (TH) (now integral part of the Karlsruhe Institute of Technology (KIT)). The financial and intellectual support is gratefully acknowledged. I want to express my gratitude to a couple of people who stimulated and encouraged me in recent years. First of all, I am much obliged to my advisor Prof. Dr. Andreas Kirsch for many valuable discussions and hints during the work on this thesis. I would like to thank as well PD Dr. Frank Hettlich for kindly agreeing to be the co-examiner of the thesis. Moreover, I want to thank my present and former colleagues from the Research Training Group and from the Workgroup on Inverse Problems for a convenient working atmosphere. Especially, I appreciate diverse discussions with Dr. Armin Lechleiter, Dipl.-Math. Thomas Gauss, Dr. Arne Schneck, and Dr. Tomáš Dohnal. I also want to mention a long conversation with Prof. Dr. Kurt Busch, which helped me to consider my work in a bigger and physical context. Finally, I want to thank my mother for constant support on my way and hours of listening.

Contents

1 Introduction **1**

1.1 Physical background . . . 1

1.2 Basic notation and terminology . . . 2

1.3 Previous results and aim of this work . . . 4

2 Geometric setting and basic tools **9**

2.1 The geometric setting . . . 9

2.2 Basic function spaces . . . 12

2.3 Trace operators . . . 21

2.4 Green's identities . . . 22

2.4.1 Generalized trace operators and identities . . . 24

2.5 Additional function spaces . . . 27

3 The acoustic case **29**

3.1 The direct problem . . . 29

3.1.1 Problem formulation . . . 29

3.1.2 The Green's function and a representation theorem . . . 33

3.1.3 The near field operator . . . 35

3.2 The inverse problem . . . 38

3.2.1 Factorization of the near field operator . . . 39

3.2.2 The interior transmission eigenvalue problem . . . 45

3.2.3 The inner operator . . . 47

4 The electromagnetic case **51**

4.1 The direct problem . . . 51

4.1.1 Problem formulation . . . 51

4.1.2 The Green's tensor and representation theorems . . . 56

4.1.3 The near field operator . . . 59

4.2 The inverse problem . . . 60

4.2.1 Factorization of the near field operator 61
4.2.2 The interior transmission eigenvalue problem 67
4.2.3 The inner operator 70

5 Approximation of the near field operators 77
5.1 The acoustic case 77
5.2 The electromagnetic case 84

6 Reconstruction of the shape 93
6.1 Range tests 93
6.2 Range identity 97
6.3 Absorbing media 99
6.3.1 The acoustic case 99
6.3.2 The electromagnetic case 99
6.4 More general media 100
6.4.1 The acoustic case 100
6.4.2 The electromagnetic case 101
6.5 Regularization of the Factorization Method 101

7 Numerical solvers 105
7.1 Direct problem: A fast solver for the α-quasi-periodic Lippmann-Schwinger equation 106
7.1.1 The $\star$-periodic Lippmann-Schwinger equation 106
7.1.2 Trigonometric collocation 108
7.1.3 Two-grid iteration scheme 114
7.1.4 Extension to discontinuous contrasts 116
7.1.5 Simulation scheme 122
7.1.6 Numerical examples 124
7.2 Inverse problem: Reconstruction of the medium shape 134
7.2.1 Reconstruction scheme (for simulated data) 134
7.2.2 Numerical examples 135
7.3 Conclusion 143

Index 144

Bibliography 147

Chapter 1

Introduction

1.1 Physical background

Let us start with a description of the physical background of scattering problems. These problems refer to the scattering of an acoustic or electromagnetic *incident field* at some *scattering object* (or *scatterer*) embedded in some *background medium* in $\mathbb{R}^d$ for $d \in \{2,3\}$. Depending on whether or not the incident field can propagate inside the scatterer, one distinguishes between *penetrable* and *impenetrable* objects. The latter ones are often called 'obstacles'.

An acoustic field is a pressure field. The position-dependent propagation speed $c(x)$ of sound is determined by properties of the matter at x. We do not examine this relation here, but simply consider the speed of sound as the characteristic physical feature for the acoustic scattering problem. The function $n(x) = c_0^2/c(x)^2$ of the space variable x is called the *refraction index* and appears naturally in the mathematical model later on. Here, c_0 denotes the speed of sound in air. We assume the propagation speed not to depend on the direction, so that c and n are scalar-valued. For acoustic scattering, we let the background medium consist of air, so that n equals 1 there. To include the possibility of energy absorption, we let the refraction index have the general form $n = n_R + \mathrm{i}\, n_I$ with functions $n_R, n_I : \mathbb{R}^d \to \mathbb{R} \cup \{+\infty\}$ of the space variable. For an impenetrable acoustic scatterer, $|n|$ is finite only outside the scatterer.

The propagation of an electromagnetic field is governed by the position-dependent *permittivity* $\varepsilon(x)$, *conductivity* $\sigma(x)$, and *permeability* $\mu(x)$. Roughly speaking, the permittivity refers to the ability of a material to transmit an electric field and the conductivity to its ability to conduct an electric current. The permeability indicates the degree of magnetization of a material in response to a magnetic field. We combine the former two functions in the complex-valued permittivity $\hat{\varepsilon}(x) = \varepsilon(x) + \mathrm{i}\,\sigma(x)/\omega$. In the mathematical formulation of the electromagnetic problem, we use $\varepsilon_r(x) = \hat{\varepsilon}(x)/\varepsilon_0$ and $\mu_r(x) = \mu(x)/\mu_0$ where the subscript 0 denotes the respective constant values of ε and μ in vacuum. Opposed to acoustic fields, electromagnetic fields can also propagate in vacuum, and we let vacuum make up the background in this case. We confine ourselves in this work to isotropic and non-magnetic materials, so the functions above are scalar-valued and μ_r is constant equal to 1 in the whole space $\mathbb{R}^d$. If the scatterer is impenetrable, i.e. a perfect conductor, $|\varepsilon_r|$ is finite only outside the scatterer.

In both settings above, the frequency of the incident radiation is denoted by $\omega > 0$. The *wave number* k_0 is given by $k_0 = \omega/c_0$ in the acoustic and by $k_0 = \omega\sqrt{\varepsilon_0\,\mu_0}$ in the electromagnetic case. Throughout the work, we assume that the fields are *time-harmonic*, meaning that their dependence on time t is described by $\exp(-\mathrm{i}\,\omega t)$ for all $t \in \mathbb{R}$. We factor out this dependence and deal with the time-independent part only. For details about the constitutive physical laws we refer the reader to standard literature like e.g. [35]. For a thorough (mathematical) reference for electromagnetic wave theory see [56].

1.2 Basic notation and terminology

Notation

Let $\mathbb{C}^\star$ denote the set $\mathbb{C}^\star = \{c_R + \mathrm{i}c_I : c_R, c_I \in \mathbb{R} \cup \{-\infty, +\infty\}\}$. We define the so-called *contrast* $q : \mathbb{R}^d \to \mathbb{C}^\star$ for $d \in \{2,3\}$ by $q(x) = n(x) - 1$ in the acoustic and by $q(x) = 1 - 1/\varepsilon_r(x)$ in the electromagnetic case. This function naturally arises in the formulation of the scattering problems later on. We let $|q|$ take arbitrary values in the range $[0, +\infty]$, which allows us to handle penetrable and impenetrable objects in a uniform manner here. For a concise formulation of the next section, we summarize some basic notation in the following definition.

Definition 1.1. Let $f : \mathbb{R}^d \to \mathbb{C}^\star$ and e_j denote the j-th unit vector in $\mathbb{R}^d$ for $j \in \{1,\dots,d\}$, $d \in \{2,3\}$.

(i) We call the function f *periodic with period $p > 0$ in the x_j-dimension* in the usual way if

$$f(x + p\,e_j) = f(x) \qquad \text{for almost all } x \in \mathbb{R}^d.$$

(ii) Let $\Lambda = (p_1,\dots,p_d) \in \mathbb{R}^d \setminus \{0\}$ with $p_j \geq 0$ for all $j \in \{1,\dots,d\}$. If f is periodic with period p_j in the x_j-dimension for j with $p_j > 0$ and either constant, i.e. trivially periodic, almost everywhere or not at all periodic in the x_j-dimension for j with $p_j = 0$, then we call f *Λ-periodic* for short.

(iii) If we call f simply *periodic* and do not specify Λ, we assume that the following holds:

- f is periodic in the x_1-dimension with some period $p_1 > 0$,
- f is either periodic in the x_{d-1}-dimension with some period $p_{d-1} > 0$ or constant almost everywhere in the x_{d-1}-dimension.

(iv) A set $S \subseteq \mathbb{R}^d$ is called the *essential support* of f if S is the smallest closed set in $\mathbb{R}^d$ such that f is unequal to zero almost everywhere in S and vanishes almost everywhere in $\mathbb{R}^d \backslash S$. For convenience, we denote this set also by $\operatorname{supp} f$ and call it simply the *support* of f.

(v) We call a set $S \subseteq \mathbb{R}^d$ periodic with period $p > 0$ in the x_j-dimension, Λ-periodic, or periodic if its indicator function id_S has the respective property. In particular, if f is periodic with period $p > 0$ in the x_j-dimension, Λ-periodic, or periodic, then so is $\operatorname{supp} f$ respectively.

Moreover, we always stick to the following convention.

Definition 1.2. Whenever we talk about periodic contrasts q, in addition to Definition 1.1 (iii) we assume that

- $q(x) = 0$ for almost all $x \in \mathbb{R}^d$ with $x_d \geq h$ for some $h > 0$.

For simplicity, we omit in the following the annotation 'almost everywhere' in the statements that q vanishes or is constant almost everywhere in some set. Throughout the work, we exclusively consider scattering objects associated with periodic contrasts q. Further conditions on q are announced later. As an essential geometric object for our problem treatment, we define the so-called *unit cell* by

$$\begin{aligned} \Pi &= \left(-\tfrac{p_1}{2}, \tfrac{p_1}{2}\right) \times \mathbb{R} && \text{for } d = 2, \\ \Pi &= \left(-\tfrac{p_1}{2}, \tfrac{p_1}{2}\right) \times \left(-\tfrac{r_2}{2}, \tfrac{r_2}{2}\right) \times \mathbb{R} && \text{for } d = 3, \end{aligned}$$

where r_2 is set to 2π if q is constant in the x_2-direction and to $p_2 > 0$ otherwise. The restriction $q\big|_\Pi$ characterizes q everywhere except on a Lebesgue null set in $\mathbb{R}^d$. It will become clear in the next chapter that, under natural conditions, scattering problems for periodic objects can be posed and handled entirely in the unit cell. Finally, we let $\Omega' \subset \mathbb{R}^d$ be any open periodic set whose closure contains $\operatorname{supp} q$ and define

$$\Omega = \Omega' \cap \Pi, \qquad \Gamma = \partial\Omega \cap \Pi, \qquad \Omega^{\mathrm{ext}} = \Pi \backslash \overline{\Omega}. \tag{1.1}$$

To keep the formulations simple, by the 'scattering object' or 'scattering medium' we mean the pair $(q, \operatorname{supp} q)$ in the following. The context makes it clear whether the contrast or its support is addressed in particular. In the mathematical treatment, penetrable periodic objects lead to transmission problems in Π, with some transmission condition(s) set at Γ. Impenetrable objects are modeled by boundary value problems in $\Pi \setminus \operatorname{supp} q$, imposing some characteristic boundary condition(s) on the field at $\partial(\operatorname{supp} q) \cap \Pi$. The type of the condition(s) depends on the physical properties of the scattering object and on whether acoustic or electromagnetic fields are considered.

Terminology

For a first orientation, we repeat some common terminology. All scattering objects in $\mathbb{R}^d$, $d \in \{2,3\}$, which are associated with periodic contrasts q are subsumed under the term 'periodic media'. Those for which $\operatorname{supp} q$ is simply connected and q is constant in $\operatorname{supp} q$ are called '(diffraction) gratings'. Gratings in $\mathbb{R}^3$ divide into 'lamellar gratings', which are invariant in the x_2-direction, and 'crossed gratings', corresponding to $p_2 > 0$. If a grating is impenetrable and a Dirichlet boundary condition is imposed on the total field at $\partial(\operatorname{supp} q) \cap \Pi$, the grating is said to be 'perfectly reflecting / conducting'. In the acoustic case, this corresponds to a 'sound- or acoustically soft' grating, whereas a Neumann boundary condition is used to model a 'sound- or acoustically hard' grating, cf. [17]. The boundary of a grating is referred to as the 'interface' or 'scattering surface'. It is also called the 'profile', especially when it is given as the graph of some function. We remark that $\partial(\operatorname{supp} q) \cap \Pi$ might consist of more than one connected component. The problem to compute a scattered field for a specified medium and incidence is called the *direct problem*. Complementary, as *inverse problem* we consider the identification of the shape of the medium, i.e. the support of the contrast, by means of the scattered fields for a number of incident fields. Clearly, for a grating it is about the identification of its interface. The objective is usually not a full reconstruction of the contrast q as a function of the spatial variable x.

1.3 Previous results and aim of this work

To give an overview of the state of research in the field of wave scattering from periodic objects, we classify now some important publications. For lamellar gratings, there are two cases for scattering of an electromagnetic incident plane wave.

If the incidence direction is orthogonal to the x_2-direction, then the resulting scattered field is invariant in the x_2-coordinate. The governing equations for the electromagnetic field, the *time-harmonic Maxwell's equations*, can be reduced in this case to a system of two scalar equations for the x_2-components of the electric and the magnetic field. If the permittivity is piecewise constant in the principal x_1-x_3-plane, both equations are of the same type as the governing equation for an acoustic field, the *Helmholtz equation*, in 2D. Moreover, as an important feature of the case of an orthogonal incidence, they are not coupled by any supplemental conditions in the problem model, hence the x_2-components can be considered separately. The scattered field can then be decomposed into modes with either *transverse electric (TE)* polarization, corresponding to a vanishing electric x_2-component, or *transverse magnetic (TM)* polarization, where the magnetic x_2-component is zero. In other words, for transverse electric modes the amplitude vector of the electric field is perpendicular to the x_2-direction and for transverse magnetic modes so is the amplitude vector of the magnetic field. For a comprehensive study of this problem, see the monograph by WILCOX [71]. Opposed to that, an oblique incident plane wave leads to so-called *conical diffraction*. Here, the equations of the electromagnetic field can again be transformed into a system of the above form, but the equations for the x_2-components are coupled at the interface of the grating, cf. [20]. The monograph [57], edited by PETIT and published about thirty years ago, can serve as an entry point to the whole subject of scattering from gratings. The authors discuss many aspects for different types of gratings, including the above-mentioned. There are various widely-used numerical methods for these problems, among which are powerful boundary integral equation techniques, cf., e.g., [58, 54]. Further information can be found in Chapters 2–5 in [10]. In [37], the direct problem of scattering of a plane wave from a smooth sound-soft grating in 2D is treated. The author establishes existence of a solution for all frequencies and uniqueness for all but finitely many frequencies. The exceptional ones are the so-called *Rayleigh frequencies*. The articles [38, 23, 22] deal with a closely related inverse problem for a perfectly reflecting grating in 2D. In [38] it is proven that, except for the Rayleigh frequencies, the scattered fields which belong to all so-called 'quasi-periodic' incident fields and are measured on a straight line above the grating uniquely determine a smooth profile. The more recent paper [22] establishes the same result for piecewise linear profiles by means of significantly less data. The survey [9] summarizes the status of research in 2003 concerning the inverse problem for perfectly reflecting gratings in 2D and 3D and provides a rich bibliography of related publications.

Results on the uniqueness of the inverse problem in the 2D case of a penetrable grating with Lipschitz profile are obtained in [24], extending the approach followed in [31] for a smooth, perfectly reflecting grating. This work deals with the direct and the inverse problem for penetrable periodic media of finite height in $\mathbb{R}^3$, with variable material parameter n and ε_r, respectively. We require that the material parameter has essentially finite absolute value in Ω' and that the contrast satisfies $q(x) = 0$ for almost all $x \in \mathbb{R}^3$ with $x_3 \leq -h$ for some $h > 0$, in addition to the conditions in Definitions 1.1 and 1.2. The medium is allowed to be disconnected and is irradiated by some time-harmonic acoustic or electromagnetic field. For the treatment of the inverse problem, we will make some additional requirements. Earlier works in this direction include [19, 8, 63, 64] for the direct problem of electromagnetic scattering from general periodic media in 3D and [5, 4] for the inverse problem of reconstructing impenetrable lamellar gratings. In [8], the same type of periodic media is considered as in this work. The author chooses a variational approach to prove existence and uniqueness of a solution to the electromagnetic direct problem (in an appropriate sense) for a plane wave incidence, under exclusion of the Rayleigh frequencies. The article [63] extends the results for the direct problem obtained in [21] for periodic contrasts which are piecewise constant and invariant in one direction to a large class of periodic contrasts, including those treated here. The related work [64] also addresses conical diffraction from those of the general media which are constant in one direction. The inverse problem for a smooth and perfectly reflecting grating is solved for all but the Rayleigh frequencies in [5, 4], by setting up and applying a suitable *Factorization Method*. This method belongs to the class of qualitative reconstruction methods (cf. [15]) and has been proposed by KIRSCH in the context of scattering from bounded objects in [39, 40]. It has a complete and profound mathematical foundation and initiated vigorous research in the field, cf. the monograph by KIRSCH AND GRINBERG [45] and the references given therein. In the main part of our work, we develop a variant of this method to solve the inverse problem for general, in particular inhomogeneous periodic media. Complementing the theoretical framework, in the final chapter we setup numerical solvers for the direct as well as the inverse acoustic problem in 2D. For the direct problem, we follow and enhance the approach by VAINIKKO in [68], cf. also the monograph by VAINIKKO AND SARANEN [61]. This leads to an efficient solver for the so-called Lippmann-Schwinger equation, a Fredholm volume integral equation of the second kind, which is proven to provide an equivalent formulation of the direct problem. Our reconstruction scheme for the inverse problem describes the implementation of the central assertion of the Factorization Method, which is established in Chapter 6. We combine and apply both solvers in a couple of numerical examples to demonstrate the applicability of our results.

Chapter 2

Geometric setting and basic tools

2.1 The geometric setting

Throughout the work, we consider scattering problems for periodic media in $\mathbb{R}^3$. We recall that the contrast q is defined by $q(x) = n(x) - 1$ in the acoustic and by $q(x) = 1 - 1/\varepsilon_r(x)$ in the electromagnetic case, where n is the complex-valued refraction index and ε_r is the complex-valued relative permittivity. The scattering medium is embedded in a homogeneous background matter which occupies the whole space $\mathbb{R}^3$ and acts like air in acoustics and like vacuum in electromagnetics, respectively. An open periodic set $\Omega' \subset \mathbb{R}^3$ with finite extension in the x_3-direction is chosen such that $\operatorname{supp} q \subseteq \overline{\Omega'}$. By a proper scaling, we can guarantee without loss of generality that q is 2π-periodic in the x_1- as well as the x_2-direction, so $\Lambda = (2\pi, 2\pi, 0)^T$. Thus, the unit cell Π is given by $\Pi = (-\pi, \pi)^2 \times \mathbb{R}$. Moreover, we require that Ω' as well as $\Omega = \Omega' \cap \Pi$ are Lipschitz and $\operatorname{supp} q \cap \overline{\Pi}$ is not degenerate, according to the following definition.

Definition 2.1. Let $S \subseteq \mathbb{R}^3$.

(i) We call S degenerate if $\partial(S^\circ) \neq \partial S$, where as usual S° denotes the interior of S and $\partial S = \overline{S} \setminus S^\circ$ denotes the boundary of S.

(ii) We call S a *Lipschitz set* or just *Lipschitz* if S is open and if for every $x \in \partial S$ there exists a neighborhood $\mathcal{O} \subset \mathbb{R}^3$ of x and a new orthogonal coordinate system with coordinates $(y_1, y_2, y_3) = y$, resulting from the original coordinate system by a rotation plus a translation, such that

   (a) $\mathcal{O}$ is a cube in the new coordinates, i.e. $\mathcal{O} = \{y : -a_j < y_j < a_j,\ j = 1, 2, 3\}$ for some vector $a \in (\mathbb{R}^+)^3$.

   (b) There exists a scalar-valued Lipschitz continuous function τ defined in

$$\mathcal{O}' = \{(y_1, y_2) : -a_j < y_j < a_j,\ j = 1, 2\}$$

that satisfies $|\tau(y')| \le a_3/2$ for all $y' \in \mathcal{O}'$ and

$$S \cap \mathcal{O} = \{y \in \mathcal{O} : y_3 < \tau((y_1,y_2))\},$$
$$\partial S \cap \mathcal{O} = \{y \in \mathcal{O} : y_3 = \tau((y_1,y_2))\}.$$

We note that, in particular, S might be unbounded and disconnected. S is called a *Lipschitz domain* if S is Lipschitz and connected.

Further restrictions on Ω and q are stated in the places where they are needed. The medium is irradiated by a time-harmonic acoustic or electromagnetic field of the form

$$U(x,t) = \mathrm{Re}\{u(x)\,e^{-\mathrm{i}\omega t}\}$$

with fixed frequency $\omega > 0$ and a complex-valued, space-dependent part u. In acoustics, U is scalar-valued and represents a pressure field, whereas in electromagnetics it is vector-valued and might either represent the electric or the magnetic field. In general, the space-dependent part u does not share the Λ-periodicity of the contrast q. In fact, we consider so-called *α-quasi-periodic* fields $u_\alpha : \mathbb{R}^3 \to \mathbb{C}^d$, $d \in \{1,3\}$, as defined next.

Definition 2.2. Let $\alpha \in \mathbb{R}^2 \times \{0\}$. A function $u_\alpha : \mathbb{R}^3 \to \mathbb{C}^d$, $d \in \{1,3\}$, which satisfies

$$u_\alpha(x+\Lambda \odot e_j) = e^{\mathrm{i}\alpha_j\Lambda_j}\,u_\alpha(x) \qquad \text{for all } j \in \{1,2,3\} \text{ and almost all } x \in \mathbb{R}^3$$

is called *α-quasi-periodic* with *quasi-period* Λ and *phase shift* α. Here, $\odot$ denotes the componentwise multiplication and e_j again the j-th unit vector. For such a function u_α,

$$\widetilde{u}(x) = e^{-\mathrm{i}\alpha\cdot x}\,u_\alpha(x)$$

is the Λ-periodic counterpart.

In the following, we omit the indication of the quasi-period Λ. We note that for $\alpha = (0,0,0)^T$, an α-quasi-periodic function is Λ-periodic. Since $\Lambda = (2\pi,2\pi,0)^T$ is fixed, we call Λ-periodic functions just periodic. Moreover, we abbreviate $\alpha = (0,0,0)^T$ by $\alpha = 0$ from now on. Let us shortly illustrate the notion of α-quasi-periodicity for the common choice of a plane wave incidence. The space-dependent part $u^{(\mathrm{pw})} : \mathbb{R}^3 \to \mathbb{C}^d$ of a plane wave with $d \in \{1,3\}$ is given as

$$u^{(\mathrm{pw})}(x) = p\,e^{\mathrm{i}k\theta\cdot x}. \tag{2.1}$$

Here, k is the wave number, $\theta \in \mathbb{R}^3$ is a unit vector which indicates the propagation direction of the wave, and $p \in \mathbb{R}^d \backslash \{0\}$ is either the scalar amplitude of

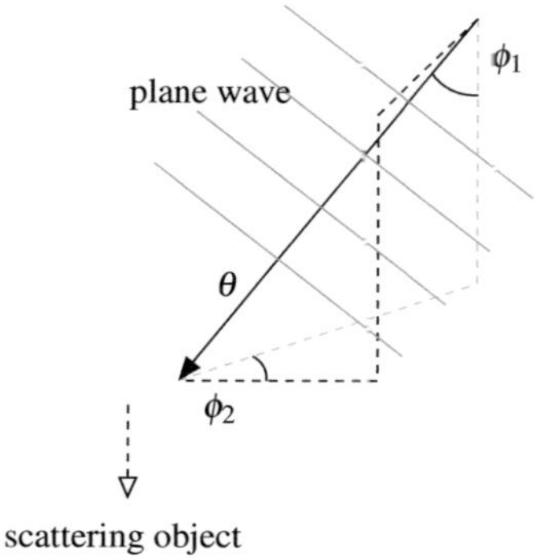

Figure 2.1: Plane wave incidence

an acoustic wave ($d = 1$) or the polarization of an electric or magnetic transverse wave, satisfying $p \cdot \theta = 0$ ($d = 3$). The direction θ can be stated in terms of the incidence angles as

$$\theta = (\sin\phi_1 \cos\phi_2, \sin\phi_1 \sin\phi_2, -\cos\phi_1)^T,$$

where $\phi_1 \in [0, \pi/2)$ and $\phi_2 \in [0, 2\pi)$. Figure 2.1 sketches the geometric situation for a plane wave. Obviously, $u^{(\mathrm{pw})}$ is α-quasi-periodic with

$$\alpha = k(\sin\phi_1 \cos\phi_2, \sin\phi_1 \sin\phi_2, 0)^T$$

and $e^{-\mathrm{i}\,\alpha\cdot x}u^{(\mathrm{pw})} = p\,e^{-\mathrm{i}k\cos\phi_1 x_3}$ is trivially periodic. However, in the main part of this work we consider a broader class of α-quasi-periodic incident fields, made up by weighted superpositions of the fields of point sources on some flat surface. The point sources are either acoustic point sources or magnetic dipoles. In this class of fields, a plane wave (precisely, its space-dependent part) can be approximated based on Huygens' Principle. In addition to (1.1), we use the notation

$$\begin{aligned} \Gamma_\pm &= \{x \in \Pi : x_3 = m_\pm\}, \\ \Omega_\pm &= \{x \in \Pi : x \text{ is connected to } \Gamma_\pm \text{ in } \Omega^{\mathrm{ext}} = \Pi \backslash \overline{\Omega}\}, \\ R_\pm &= \{x \in \Pi : x_3 \gtrless m_\pm\} \end{aligned} \tag{2.2}$$

where $m_+ > \sup\{x_3 : x \in \Omega'\}$ and $m_- < \inf\{x_3 : x \in \Omega'\}$. Without loss of generality, we assume that $\sup\{x_3 : x \in \Omega'\} \geq 0$ and $\inf\{x_3 : x \in \Omega'\} \leq 0$. Note that the sets Ω_+ and Ω_- might coincide if Ω' is disconnected. Figure 2.2 illustrates the setting apart from Ω^{ext} and $\Omega_\pm$.

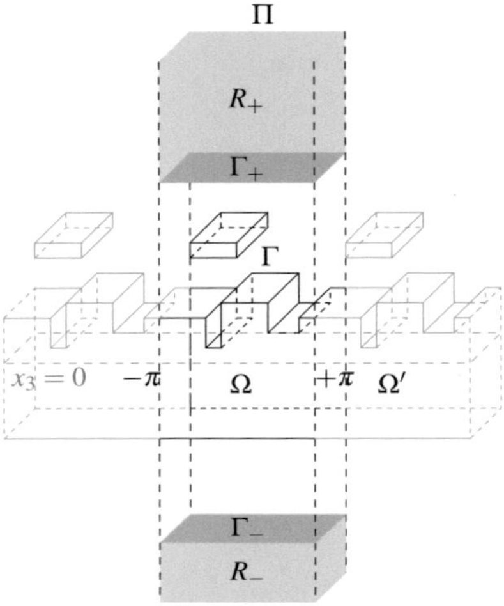

Figure 2.2: The geometric setting

2.2 Basic function spaces

Clearly, a central role in the problem treatment is played by the proper choice of function spaces. We mostly work with *Sobolev spaces* on Lipschitz sets and their boundaries. In this section, we introduce these spaces after summarizing some basic definitions.

Definition 2.3. Let $S \subseteq \mathbb{R}^3$ be some open set and φ_j, $j \in \mathbb{N}$, be a sequence in $C^\infty(S) = \bigcap_{r \geq 0} C^r(S)$. Let K denote a compact subset of S in the following. We define the function spaces

$$C_K^r(S) = \{u \in C^r(S) : \operatorname{supp}(u) \subseteq K\} \qquad \text{and} \qquad C_K^\infty(S) = \bigcap_{r \geq 0} C_K^r(S)$$

and also $C_0^\infty(S) = \{u : u \in C_K^\infty(S) \text{ for some } K\}$. Let now $\mu \in \mathbb{N}_0^3$ be a multi-index of order $|\mu| = \sum_{j=1}^3 \mu_j$ and

$$\partial^\mu u(x) = \frac{\partial^{\mu_1}}{\partial^{\mu_1} x_1} \frac{\partial^{\mu_2}}{\partial^{\mu_2} x_2} \frac{\partial^{\mu_3}}{\partial^{\mu_3} x_3} u(x) \tag{2.3}$$

denote the partial derivative of $u \in C^{|\mu|}(S)$ with respect to the index μ. We write

$$\varphi_j \to 0 \text{ in } C_K^\infty(S) \quad \Longleftrightarrow \quad \partial^\mu \varphi_j \to 0 \text{ uniformly in } K \text{ for all } \mu \in \mathbb{N}_0^3 \tag{2.4}$$

and similarly

$$\varphi_j \to 0 \text{ in } C_0^\infty(S) \quad \Longleftrightarrow \quad \varphi_j \to 0 \text{ in } C_K^\infty(S) \text{ for some } K. \tag{2.5}$$

By $\mathcal{E}(S)$ we denote the space $C^\infty(S)$ combined with the notion of sequential convergence in the sense that

$$\varphi_j \to 0 \text{ in } \mathcal{E}(S) \quad \Longleftrightarrow \quad \partial^\mu \varphi_j \to 0 \text{ uniformly in } K \text{ for all } \mu \in \mathbb{N}_0^3 \text{ and all } K.$$

The space $\mathcal{E}(\overline{S})$ consists of all functions $u \in \mathcal{E}(S)$ for which all derivatives have continuous extensions to ∂S. Moreover, $\mathcal{D}(S)$ denotes the space $C_0^\infty(S)$ with sequential convergence in the sense of (2.5). The elements of $\mathcal{D}(S)$ are referred to as *test functions* on S. A linear, sequentially continuous functional $l : \mathcal{D}(S) \to \mathbb{C}$ is called a *(Schwartz) distribution* on S. The set of all distributions on S is denoted by $\mathcal{D}(S)'$.

Sobolev spaces of integer order and their duals

We can now continue to define the notion of a *weak (partial) derivative* and of a *Sobolev space (of integer order)*. In the following, we let $S \subseteq \mathbb{R}^3$ be some Lipschitz set.

Definition 2.4. Let $u \in L^2(S)$ and $\mu \in \mathbb{N}_0^3$ be fixed. If there is a function $f_\mu \in L^2(S)$ such that

$$\langle u, \partial^\mu \phi \rangle_S = (-1)^{|\mu|} \langle f_\mu, \phi \rangle_S \qquad \text{for all } \phi \in \mathcal{D}(S),$$

then f_μ is called the *weak partial derivative* of u *with respect to the index* μ.

We just remark here that if $u \in L^2(S)$ has a classical partial derivative, then the corresponding weak partial derivative exists and coincides with the classical one. For this reason, we denote the function f_μ, in generalization of (2.3), also by $\partial^\mu u$. For $m \in \mathbb{N}_0$, the space

$$H^m(S) = \{u \in L^2(S) : \partial^\mu u \in L^2(S) \text{ for all } \mu \text{ with } |\mu| \le m\} \tag{2.6}$$

is called the *Sobolev space* of *order m* on S. Equipped with the inner product

$$\langle u, v \rangle_{H^m(S)} = \sum_{|\mu| \le m} \int_S \partial^\mu u \, \overline{\partial^\mu v} \, \mathrm{d}x, \tag{2.7}$$

$H^m(S)$ is a Hilbert space with the naturally induced norm $\|\cdot\|_{H^m(S)}$. An alternative characterization of $H^m(S)$ is given by the closure

$$H^m(S) = \overline{\mathcal{E}(\overline{S})}^{\|\cdot\|_{H^m(S)}}. \tag{2.8}$$

It is an important observation that for bounded S and $m \geq 1$ the space $\mathcal{D}(S)$ is not dense in $H^m(S)$, thus we define in addition

$$H_0^m(S) = \overline{\mathcal{D}(S)}^{\|\cdot\|_{H^m(S)}}.$$

For $m = 0$, there holds $H^0(S) = H_0^0(S) = L^2(S)$, cf. Corollary 3.5 in [53]. Moreover, for any Lipschitz set $S \subseteq \mathbb{R}^3$, we define the space

$$H_{\mathrm{loc}}^m(S) = \{u : S \to \mathbb{C} : u|_{\widetilde{S}} \in H^m(\widetilde{S}) \text{ for any compact } \widetilde{S} \subset S\}. \tag{2.9}$$

As a particular case (for $m = 0$), we have

$$L_{\mathrm{loc}}^2(S) = \{u : S \to \mathbb{C} : u|_{\widetilde{S}} \in L^2(\widetilde{S}) \text{ for any compact } \widetilde{S} \subset S\}.$$

Finally, by $H^{-m}(S)$ with $m \in \mathbb{N}_0$ we denote the space of distributions $u \in \mathcal{D}(S)'$ that admit a representation of the form

$$u = \sum_{|\mu| \leq m} \partial^\mu g_\mu \qquad \text{with } g_\mu \in L^2(S),$$

see Lemma 1.2 in [28]. For bounded S, $H^{-m}(S)$ is the (topological) dual space of $H_0^m(S)$, rather than that of $H^m(S)$, see Subsection 6.4.9 in [59]. For $S = \mathbb{R}^3$, it is the dual of $H^m(S)$.

Sobolev spaces of fractional order

In addition to the Sobolev spaces of integer order, we can define such of fractional order. To this end, let $s > 0$ be non-integer and $s = m + \sigma$ be its unique decomposition with $m \in \mathbb{N}_0$ and $\sigma \in (0,1)$. The semi-norm $|\cdot|_\sigma$ induced by the Hermitian form

$$(u,v)_\sigma = \int_S \int_S \frac{(u(x) - u(y))\overline{(v(x) - v(y))}}{|x-y|^{3+2\sigma}} \,\mathrm{d}x\,\mathrm{d}y$$

is called the *Slobodeckiĭ semi-norm*. Using this, we define the Sobolev space $H^s(S)$ of fractional order s by

$$H^s(S) = \{u \in H^m(S) : |u|_\sigma < \infty\}. \tag{2.10}$$

Equipped with the inner product

$$\langle u,v \rangle_{H^s(S)} = \langle u,v \rangle_{H^m(S)} + \sum_{|\mu|=m} (\partial^\mu u, \partial^\mu v)_\sigma, \tag{2.11}$$

$H^s(S)$ is a Hilbert space with the naturally induced norm $\|\cdot\|_{H^s(S)}$. It can also be characterized by

$$H^s(S) = \overline{\mathcal{E}(\overline{S})}^{\|\cdot\|_{H^s(S)}}. \tag{2.12}$$

Analog to $H_0^m(S)$, we define the space

$$H_0^s(S) = \overline{\mathcal{D}(S)}^{\|\cdot\|_{H^s(S)}}.$$

The scalars m from (2.6) and s from (2.10) are sometimes called the *Sobolev indices* of the corresponding spaces.

α-quasi-periodic Sobolev spaces and their duals

Since we are mainly interested in α-quasi-periodic functions (with quasi-period $\Lambda = (2\pi, 2\pi, 0)^T$), the following spaces are especially important to us. Let here $S \subseteq \Pi$ be a Lipschitz set in the unit cell Π and $s > 1/2$. We define the space

$$\mathcal{E}_\alpha(\overline{S}) = \left\{u_\alpha : u_\alpha = U_\alpha\big|_S \text{ for some } \alpha\text{-quasi-periodic } U_\alpha \in \mathcal{E}(\mathbb{R}^3)\right\}. \tag{2.13}$$

Then, in analogy to (2.12), we set

$$H_\alpha^s(S) = \overline{\mathcal{E}_\alpha(\overline{S})}^{\|\cdot\|_{H^s(S)}}. \tag{2.14}$$

Equipped with the inner product (2.7) or (2.11), depending on whether s is integer or not, $H_\alpha^s(S)$ is a Hilbert space. We note that if $S \neq \Pi$, in particular if $\partial S \cap \partial \Pi = \emptyset$, then for every $w_\alpha \in H_\alpha^s(S)$ there are $u_\alpha, v_\alpha \in H_\alpha^s(\Pi)$ with $u_\alpha \not\equiv v_\alpha$ and $w_\alpha = u_\alpha|_S = v_\alpha|_S$. Now, we make some remarks concerning the dual space of $H_\alpha^s(S)$. First, for $\alpha = 0$ any function $\widetilde{u} \in \mathcal{E}_{\alpha=0}(\overline{S})$ is the restriction to S of a periodic function $\widetilde{U}$ on $\mathbb{R}^3$. Hence, for every fixed x_3, the function $\widetilde{U}_{x_3}(x_1, x_2) = \widetilde{U}(x)$ is a $(2\pi, 2\pi)$-periodic function on $\mathbb{R}^2$. Every such function can be interpreted as to live on the 2-torus $\mathbb{T}^2 = S^1 \times S^1$, which is the direct product of two unit circles S^1, see Figure 2.3. These circles reflect the periods in the x_1-direction and the x_2-direction, respectively. Now, let $\Pi_j \subset \partial\Pi$ for $j = 1, 2, 3, 4$ be the four closed faces of $\partial\Pi$ given by

$$\Pi_1 = \{-\pi\} \times [-\pi, \pi] \times \mathbb{R}, \qquad \Pi_2 = \{\pi\} \times [-\pi, \pi] \times \mathbb{R},$$
$$\Pi_3 = [-\pi, \pi] \times \{-\pi\} \times \mathbb{R}, \qquad \Pi_4 = [-\pi, \pi] \times \{\pi\} \times \mathbb{R}.$$

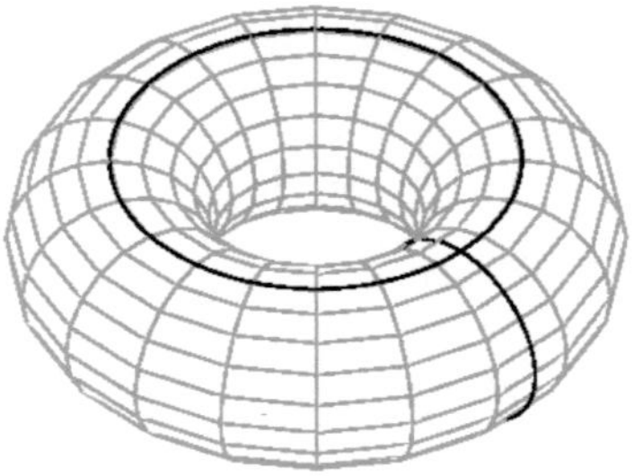

Figure 2.3: the torus $\mathbb{T}^2$

Having this, we let

$$\begin{aligned}
\Gamma_{1,2} &= \{x \in \Pi_1 : x \in (\partial S \cap \Pi_1)^\circ \wedge x + (2\pi, 0, 0)^T \in (\partial S \cap \Pi_2)^\circ\}, \\
\Gamma_{3,4} &= \{x \in \Pi_3 : x \in (\partial S \cap \Pi_3)^\circ \wedge x + (0, 2\pi, 0)^T \in (\partial S \cap \Pi_4)^\circ\}, \\
\Gamma_1 &= \Gamma_{1,2} \cup \{x + (2\pi, 0, 0)^T : x \in \Gamma_{1,2}\}, \\
\Gamma_2 &= \Gamma_{3,4} \cup \{x + (0, 2\pi, 0)^T : x \in \Gamma_{3,4}\}, \\
\widehat{\Gamma} &= \partial S \backslash (\Gamma_1 \cup \Gamma_2).
\end{aligned}$$

These sets are defined simply in order to separate the 'periodic part' of a function in $H^s_{\alpha=0}(S)$. The situation in 2D is illustrated in Figure 2.4. Precisely, for any sufficiently regular periodic function $\widetilde{u}$, we find

$$\int_{\partial S} \frac{\partial \widetilde{u}}{\partial \nu} \,\mathrm{ds} = \int_{\widehat{\Gamma}} \frac{\partial \widetilde{u}}{\partial \nu} \,\mathrm{ds},$$

where ν denotes the outward unit normal vector to S. With

$$\mathcal{E}_\circ(\overline{S}) = \left\{\widetilde{u} \in \mathcal{E}_{\alpha=0}(\overline{S}) : \partial^\mu \widetilde{u}\big|_{\widehat{\Gamma}^\circ} = 0 \text{ for all } \mu \in \mathbb{N}_0^3\right\},$$

by $H^{-s}_{\alpha=0}(S)$ we denote the dual space of

$$H^s_\circ(S) = \overline{\mathcal{E}_\circ(\overline{S})}^{\|\cdot\|_{H^s(S)}},$$

cp. definition (2.14) and Definition 6.111 in [59]. In particular, for $S = \Pi$ it is $H^s_\circ(S) = H^s_{\alpha=0}(S)$ since $\widehat{\Gamma}^\circ = \emptyset$ in this case. On the opposite, for $\Gamma_1 = \Gamma_2 = \emptyset$ there holds $\mathcal{E}_\circ(\overline{S}) = \mathcal{D}(S)$, and so $H^{-s}_{\alpha=0}(S)$ is the dual space of $H^s_0(S)$. For $\alpha \neq 0$, by $H^{-s}_\alpha(S)$ we always denote the dual space of $H^s_0(S)$. However, at the end of

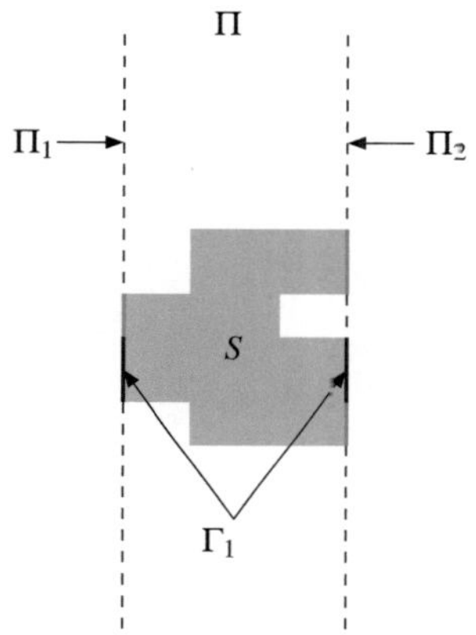

Figure 2.4: separation of the 'periodic part' in 2D

this section, we will show that there is a one-to-one correspondence between the functionals in $H^{-s}_{\alpha=\beta}(S)$ with $\beta \neq 0$ and those in $H^{-s}_{\alpha=0}(S)$. We finally consider the periodic extension S_{per} of $S \subseteq \Pi$, given by

$$S_{\text{per}} = \overset{\circ}{\overline{\{x \in \mathbb{R}^3 : x + \Lambda \odot z \in S \text{ for some } z \in \mathbb{Z}^3\}}}, \tag{2.15}$$

where again $\odot$ denotes the componentwise multiplication. Note that in general there holds $\overline{S_{\text{per}}} \cap \partial\Pi \neq \overline{S} \cap \partial\Pi$, as demonstrated by Figure 2.5. For this reason, we define $S_c = \overline{S_{\text{per}}} \cap \overline{\Pi}$, which satisfies $\overline{S_{\text{per}}} \cap \partial\Pi = S_c \cap \partial\Pi$. We remark that S_c is not a topological domain since it is closed and possibly disconnected. It might also be degenerate in the sense of Definition 2.1, as it is the case for the example shown in Figure 2.5. We will frequently consider integrals over

$$(\overline{S_{\text{per}}} \cap \partial\Pi) \cup (\partial S_{\text{per}} \cap \Pi) = (S_c \cap \partial\Pi) \cup \partial S$$

for some $S \subseteq \Pi$. Then, for any sufficiently regular periodic function $\widetilde{u}$ we obtain

$$\int_{(S_c \cap \partial\Pi) \cup \partial S} \frac{\partial \widetilde{u}}{\partial \nu} \,\text{ds} = \int_{\partial S \cap \Pi} \frac{\partial \widetilde{u}}{\partial \nu} \,\text{ds},$$

i.e. the integral contributions of $\partial\widetilde{u}/\partial\nu$ over the faces of $S_c \cap \partial\Pi$ cancel out. We want to point out that if S_c is not degenerate, then the above boils down to $(S_c \cap \partial\Pi) \cup \partial S = \partial S$ and

$$\int_{\partial S} \frac{\partial \widetilde{u}}{\partial \nu} \,\text{ds} = \int_{\partial S \cap \Pi} \frac{\partial \widetilde{u}}{\partial \nu} \,\text{ds}. \tag{2.16}$$

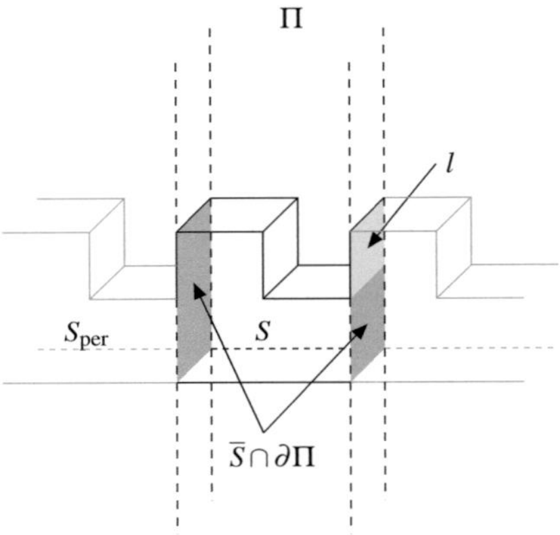

Figure 2.5: $\overline{S_{\text{per}}} \cap \partial\Pi = (\overline{S} \cap \partial\Pi) \cup l$

In particular, with the assumptions made at the beginning of this chapter, this is the case for $S = (\operatorname{supp} q \cap \overline{\Pi})^\circ$.

Vectorial Sobolev spaces

The definition of vectorial Sobolev spaces $H^s(S,\mathbb{C}^3)$ and their variants follows the above definitions in a straightforward manner. For the sake of brevity, we skip the details here and give McLean [53] as a reference. For electromagnetic scattering problems, the proper solution space, i.e. the space of finite-energy solutions, turns out to be

$$H(\operatorname{curl},S) = \{u \in L^2(S,\mathbb{C}^3) : \operatorname{curl} u \in L^2(S,\mathbb{C}^3)\},$$

normed by the graph norm

$$\|u\|_{H(\operatorname{curl},S)} = \left(\|u\|_{L^2(S,\mathbb{C}^3)} + \|\operatorname{curl} u\|_{L^2(S,\mathbb{C}^3)}\right)^{1/2},$$

with S any Lipschitz set in $\mathbb{R}^3$. Here, curl denotes the vectorial differential operator $\operatorname{curl} = \nabla\times$ with $\nabla = (\partial/\partial x_1, \partial/\partial x_2, \partial/\partial x_3)$, meant in the weak sense. Moreover, we will need the space

$$H(\operatorname{div},S) = \{u \in L^2(S,\mathbb{C}^3) : \operatorname{div} u \in L^2(S)\},$$

which is naturally normed by

$$\|u\|_{H(\operatorname{div},S)} = \left(\|u\|_{L^2(S,\mathbb{C}^3)} + \|\operatorname{div} u\|_{L^2(S)}\right)^{1/2},$$

where div denotes the weak vectorial differential operator $\operatorname{div} = \nabla\cdot$. For a bounded Lipschitz domain $S \subset \mathbb{R}^3$, it is proven in [55], Theorems 3.22 and 3.26, that the spaces $H(\operatorname{curl}, S)$ and $H(\operatorname{div}, S)$ can be characterized alternatively as the closure of $\mathcal{E}(\overline{S})$ in the norm $\|\cdot\|_{H(\operatorname{curl},S)}$ and $\|\cdot\|_{H(\operatorname{div},S)}$, respectively. In analogy to (2.16), we note that for any Lipschitz $S \subseteq \Pi$ with non-degenerate $S_c = \overline{S_{\text{per}}} \cap \overline{\Pi}$ and for any sufficiently regular periodic vectorial function $\widetilde{u}$ there hold the identities

$$\int_{\partial S} \nu \cdot \widetilde{u} \, \mathrm{ds} = \int_{\partial S \cap \Pi} \nu \cdot \widetilde{u} \, \mathrm{ds} \qquad \text{and} \qquad \int_{\partial S} \nu \times \widetilde{u} \, \mathrm{ds} = \int_{\partial S \cap \Pi} \nu \times \widetilde{u} \, \mathrm{ds}.$$

Trace spaces

It remains now to clarify whether and in which sense the concept of the restriction to ∂S of a continuous function on S with continuous extension to ∂S can be transferred to a function from one of the above Sobolev spaces. In fact, there is a consistent generalization, which leads to the concept of a *trace* of a function on ∂S. This is an essential issue in the handling of boundary value problems where Sobolev spaces serve as solution spaces. The characteristic objects in this context are the *trace operators*, which we introduce and discuss in the following sections. We will show which functions have a trace on ∂S and how the associated *trace spaces* look like. Thus, we reproduce here only a basic characterization of a trace space, cp. Subsection 3.2.1 in [55]. Let here $S \subset \mathbb{R}^3$ be some bounded Lipschitz domain and associate every $x \in \partial S$ with a set $\mathcal{O} \subset \mathbb{R}^3$ and a Lipschitz continuous function τ defined on $\mathcal{O}' = \{(y_1, y_2) : y \in \mathcal{O}\}$ which are of the type specified in Definition 2.1. Moreover, define (locally) the mapping T by $T(y') = (y', \tau(y'))$, $y' \in \mathcal{O}'$. We note that T^{-1} exists and is Lipschitz continuous on the range of T. Then a distribution u on ∂S belongs to $H^s(\partial S)$ for $|s| \leq 1$ if for all $\mathcal{O}$ and T meeting the above conditions there holds

$$u \circ T \in H^s(\mathcal{O}' \cap T^{-1}(\partial S \cap \mathcal{O})).$$

For $s \in [0, 1)$, $H^s(\partial S)$ is a Hilbert space with the inner product

$$\langle u, v \rangle_{H^s(\partial S)} = \int_{\partial S} u(x) \overline{v(x)} \, \mathrm{ds}(x) + \int_{\partial S} \int_{\partial S} \frac{(u(x) - u(y))(\overline{v(x) - v(y)})}{|x - y|^{2+2s}} \, \mathrm{d}x \, \mathrm{d}y$$

for $u, v \in H^s(\partial S)$. We have not yet characterized the dual space of a trace space. Since this is a quite technical issue and we want to confine the presentation to some basic ideas here, we refer the reader to Chapter 3 in [53]. In the acoustic case, we use the trace space for functions in $H^1(S)$, which is $H^{1/2}(\partial S)$, and its dual $H^{-1/2}(\partial S)$. An appropriate notion of traces of fields in $H(\mathrm{curl}, S)$ and the corresponding trace spaces in the electromagnetic case require a separate consideration. We postpone the details to the discussions in Section 2.3 and Subsection 2.4.1.

Periodic and α-quasi-periodic functions which have a well-defined trace on $\partial\Pi$ are determined by their behavior in the open unit cell Π. Therefore, to avoid an overload of notation, we identify such functions with their restrictions to Π. For any normed space X, we finally introduce the *dual pairing* $[\cdot,\cdot]_X : X \times X' \to \mathbb{C}$ by

$$[u, f]_X = f(u) \qquad \text{for all } u \in X,\, f \in X',$$

where, as usual, X' denotes the topological dual space of X. If X is a (complex) Hilbert space with scalar product $\langle\cdot,\cdot\rangle_X$, then by the Riesz representation theorem the mapping $\Phi : X \to X'$, $y \mapsto \langle\cdot, y\rangle_X$, is an isometric (anti-)isomorphism. Thus, for any $f \in X'$ there is some $y_f \in X$ such that $f = \langle\cdot, y_f\rangle_X$. Then there holds

$$[u, f]_X = f(u) = \langle u, y_f\rangle_X = \int u \cdot \overline{y_f}\,\mathrm{d}x \qquad \text{for all } u \in X.$$

For convenience, we will adapt later on the integral notation for any dual pairing $[\cdot,\cdot]_X$ with a normed space X and write

$$[u, f]_X = \int u \cdot f\,\mathrm{d}x \qquad \text{for all } u \in X,\, f \in X'.$$

We close the section by pointing out that there is a one-to-one correspondence between the functionals in $H^{-s}_{\alpha=\beta}(S)$ with $\beta \neq 0$ and those in $H^{-s}_{\alpha=0}(S)$. This can be seen as follows (based on [1]). According to Definition 2.2, for any $\beta \in \mathbb{R}^3$ the multiplication operator $M_\beta : H^s_{\alpha=0}(S) \to H^s_{\alpha=\beta}(S)$ defined by

$$(M_\beta u_0)(x) = e^{\mathrm{i}\beta\cdot x} u_0(x), \qquad u_0 \in H^s_{\alpha=0}(S),$$

maps a periodic function to its β-quasi-periodic counterpart. This operator is linear and bounded and has a bounded inverse M_β^{-1}. Now, denoting for short $H^s_\alpha(S)$ by X_α, there hold the identities

$$\begin{aligned} [u_\beta, f_\beta]_{X_\beta} &= [M_\beta u_0, f_\beta]_{X_\beta} = [u_0, M_\beta{}' f_\beta]_{X_0} \qquad \text{and} \\ [u_0, f_0]_{X_0} &= [M_\beta^{-1} u_\beta, f_0]_{X_0} = [u_\beta, (M_\beta^{-1})' f_0]_{X_\beta} \end{aligned} \tag{2.17}$$

for all $u_\beta \in X_\beta$, $u_0 = M_\beta^{-1} u_\beta \in X_0$, $f_\beta \in X_\beta{}'$, and $f_0 \in X_0{}'$. Using standard notation, by $M_\beta{}' : X_\beta{}' \to X_0{}'$ we denote the normed space adjoint of M_β, given by $(M_\beta{}' f_\beta)(u_0) = f_\beta(M_\beta u_0)$ for all $f_\beta \in X_\beta{}'$, $u_0 \in X_0$. The identities (2.17) already prove the assertion. Moreover, since $X_\alpha = H^s_\alpha(S)$ is a Hilbert space, there holds

$$[M_\beta u_0, f_\beta]_{X_\beta} = \langle M_\beta u_0, y_{f_\beta} \rangle_{X_\beta} = \langle u_0, M_\beta^* y_{f_\beta} \rangle_{X_0} = \langle u_0, y_{M_\beta{}' f_\beta} \rangle_{X_0} = [u_0, M_\beta{}' f_\beta]_{X_0},$$

where $M_\beta^* = \Phi_0^{-1} M_\beta{}' \Phi_\beta : X_\beta \to X_0$ is the Hilbert space adjoint of M_β and Φ_0, Φ_β are the anti-isomorphisms of the form stated above for $X = X_0$, $X = X_\beta$, respectively. Of course, a similar relation applies to $[M_\beta^{-1} u_\beta, f_0]_{X_0}$.

Further results on Sobolev spaces are given, e.g., in the thorough references [2, 53]. Specific spaces and results will be announced later, in the places where they are needed.

2.3 Trace operators

Definition 2.5. Let $S \subset \mathbb{R}^3$ be a bounded Lipschitz set. We define the operators $\gamma_D : C^\infty(\overline{S}) \to C(\partial S)$ and $\gamma_N : C^\infty(\overline{S}) \to L^2(\partial S)$ by

$$\gamma_D : u \mapsto u\big|_{\partial S} \qquad \text{and} \qquad \gamma_N : u \mapsto \frac{\partial u}{\partial \nu} = \nu \cdot \gamma_D(\nabla u)$$

for $u \in C^\infty(\overline{S})$, where ν is the exterior unit normal vector to S.

An important result is the following.

Proposition 2.6. *Let $S \subset \mathbb{R}^3$ be a bounded Lipschitz domain. The operators γ_D and γ_N from Definition 2.5 have unique extensions to bounded linear operators*

$$\gamma_D : H^1(S) \to H^{1/2}(\partial S) \qquad \textit{and} \qquad \gamma_N : H^2(S) \to H^{1/2}(\partial S).$$

The operator γ_D is called the (Dirichlet) trace operator *for S, and γ_N the* Neumann trace operator *for S.*

A proof of the extension of γ_D is given in [53], see Theorem 3.37 therein. Concerning the extension of γ_N, see Bemerkung 2.7.5 in [62]. For the treatment of electromagnetic problems, we need some more operators.

Definition 2.7. Let $S \subset \mathbb{R}^3$ be a bounded Lipschitz domain and $L^2_t(\partial S) = \{v \in L^2(\partial S, \mathbb{C}^3) : \nu \cdot v = 0 \text{ a.e. on } \partial S\}$ be the space of tangential fields on ∂S, where

ν is the exterior unit normal vector to S. Define P_t and P_T as mappings from $L^2(\partial S, \mathbb{C}^3)$ to $L_t^2(\partial S)$ by

$$P_t : w \mapsto \nu \times w \qquad \text{and} \qquad P_T : w \mapsto (\nu \times w) \times \nu$$

for $w \in L^2(\partial S, \mathbb{C}^3)$. Then, let γ_t and γ_T be the operators from $C^\infty(\overline{S}, \mathbb{C}^3)$ to $L_t^2(\partial S)$ given by $\gamma_t : u \mapsto P_t(u|_{\partial S})$ and $\gamma_T : u \mapsto P_T(u|_{\partial S})$ for $u \in C^\infty(\overline{S}, \mathbb{C}^3)$. These operators can be extended to bounded linear operators from $H^1(S, \mathbb{C}^3)$ to $L_t^2(\partial S)$ by

$$\gamma_t = P_t \circ \gamma_D \qquad \text{and} \qquad \gamma_T = P_T \circ \gamma_D.$$

Here, $\gamma_D : H^1(S, \mathbb{C}^3) \to H^{1/2}(\partial S, \mathbb{C}^3)$ is the trace operator for vectorial functions, meant as the componentwise application of the trace operator from Definition 2.5. The operator γ_t is called the *tangential trace operator* for S, and γ_T the *tangential components trace operator* for S. Moreover, let $L_n^2(\partial S) = \{v \in L^2(\partial S, \mathbb{C}^3) : \nu \times v = 0 \text{ a.e. on } \partial S\}$ denote the space of normal fields on ∂S and define P_n as a mapping from $L^2(\partial S, \mathbb{C}^3)$ to $L_n^2(\partial S)$ by

$$P_n : w \mapsto \nu \cdot w$$

for $w \in L^2(\partial S, \mathbb{C}^3)$. The operator $\gamma_n : C^\infty(\overline{S}, \mathbb{C}^3) \to L_n^2(\partial S)$ which maps $u \in C^\infty(\overline{S}, \mathbb{C}^3)$ to $P_n(u|_{\partial S})$ can be extended to a bounded linear operator $\gamma_n : H^1(S, \mathbb{C}^3) \to L_n^2(\partial S)$ with

$$\gamma_n = P_n \circ \gamma_D,$$

where γ_D is as given above. The operator γ_n is called the *normal trace operator* for S.

In the above definitions, we have used that the normal vector ν exists almost everywhere on ∂S according to Rademacher's theorem, see Satz 2.7.1 in [62].

2.4 Green's identities

Equipped with the operators from the previous section, we can now setup some important integral identities, the Green's identities.

Theorem 2.8. *Let $S \subset \mathbb{R}^3$ be a bounded Lipschitz domain. Further, let $\Delta : H^2(S) \to L^2(S)$ denote the weak Laplacian and* curl *the weak curl operator.*

(i) For all $u \in H^2(S)$ and $v \in H^1(S)$ there applies the first Green's identity

$$\int_S \Delta u \, v \, \mathrm{d}x = \int_{\partial S} \gamma_N u \, \gamma_D v \, \mathrm{d}s - \int_S \nabla u \cdot \nabla v \, \mathrm{d}x. \tag{2.18}$$

(ii) For all $u, v \in H^2(S)$ there holds the second Green's identity

$$\int_S (\Delta u\, v - u\, \Delta v)\,\mathrm{d}x = \int_{\partial S} (\gamma_N u\, \gamma_D v - \gamma_D u\, \gamma_N v)\,\mathrm{d}s. \tag{2.19}$$

(iii) For all vectorial $u, v \in H^1(S, \mathbb{C}^3)$ there holds

$$\int_S (\operatorname{curl} u \cdot v - u \cdot \operatorname{curl} v)\,\mathrm{d}x = \int_{\partial S} \gamma_t u \cdot \gamma_D v\,\mathrm{d}s. \tag{2.20}$$

Proof.

(i) We make the following operator identifications and refer to Lemmata 4.1 and 4.2 in [53]. According to the notation in [53], we write the operator $\mathcal{P} = \Delta : H^2(S) \to L^2(S)$ as

$$\mathcal{P}u = -\sum_{j=1}^{3}\sum_{k=1}^{3} \partial_j (A_{jk}\, \partial_k u), \qquad u \in H^2(S),$$

with the coefficients $A_{jk} = -\delta_{jk}$ and ∂_j denoting the weak derivative w.r.t. the j-th coordinate, $j, k \in \{1, 2, 3\}$. The 'formal adjoint' $\mathcal{P}^*$ of $\mathcal{P}$ and the 'conormal derivatives' $\mathcal{B}_\nu$ and $\widetilde{\mathcal{B}}_\nu$ are given on $H^2(S)$ by

$$\mathcal{P}^* u = \Delta u$$

$$\text{and} \qquad \mathcal{B}_\nu u = \widetilde{\mathcal{B}}_\nu u = -\nu \cdot \gamma_D(\nabla u).$$

With $\gamma_N = -\mathcal{B}_\nu$, the identity (2.18) is recognized as a special case of the first Green's identity as stated in Lemma 4.1 in [53].

(ii) The identity (2.18) is obtained as the difference of (2.18) and its dual version implied by Lemma 4.2 in [53].

(iii) Again, we make some operator identifications and refer to Lemma 4.2 in [53]. We write the operator $\operatorname{curl} : H^1(S, \mathbb{C}^3) \to L^2(S, \mathbb{C}^3)$ as

$$\mathcal{P}u = \sum_{j=1}^{3} A_j \partial_j u, \qquad u \in H^1(S, \mathbb{C}^3),$$

with the coefficient matrices

$$A_1 = \begin{bmatrix} 0 & 0 & 0 \\ 0 & 0 & -1 \\ 0 & 1 & 0 \end{bmatrix}, \quad A_2 = \begin{bmatrix} 0 & 0 & 1 \\ 0 & 0 & 0 \\ -1 & 0 & 0 \end{bmatrix}, \quad A_3 = \begin{bmatrix} 0 & -1 & 0 \\ 1 & 0 & 0 \\ 0 & 0 & 0 \end{bmatrix},$$

and ∂_j denoting the weak derivative w.r.t. the j-th coordinate, $j \in \{1,2,3\}$. The 'conormal derivative' $\mathcal{B}_\nu$ vanishes here. With

$$\widetilde{\mathcal{B}}_j u = A_j^* u = -A_j u = -e_j \times u,$$

for the 'formal adjoint' $\mathcal{P}^*$ of $\mathcal{P}$ and the 'dual conormal derivative' $\widetilde{\mathcal{B}}_\nu$ we get, respectively,

$$\mathcal{P}^* u = -\sum_{j=1}^{3} \partial_j \widetilde{\mathcal{B}}_j u = \sum_{j=1}^{3} \partial_j A_j u = \mathcal{P} u,$$

$$\widetilde{\mathcal{B}}_\nu u = \sum_{j=1}^{3} \nu_j \gamma_D(\widetilde{\mathcal{B}}_j u) = -\sum_{j=1}^{3} \nu_j (e_j \times \gamma_D u) = -\nu \times \gamma_D u.$$

Using $\gamma_t = -\widetilde{\mathcal{B}}_\nu$ and

$$\gamma_D u \cdot \widetilde{\mathcal{B}}_\nu v = \gamma_D u \cdot (-\nu \times \gamma_D v) = \gamma_D v \cdot (\nu \times \gamma_D u) = \gamma_D v \cdot \gamma_t u,$$

the identity (2.20) is found to be a special case of the dual first Green's identity as stated in Lemma 4.2 in [53]. □

2.4.1 Generalized trace operators and identities

Now, we show that the trace operators introduced above can be generalized by means of abstract variants of the Green's identities from Theorem 2.8. First, we assume that for some $u \in H^1(S)$ and some $f \in H^{-1}(\mathbb{R}^3)$ there holds $\Delta u = f$ in S. For $u \in H^1(S)$, Δu is defined as a distribution on S via

$$(\Delta u)(v) = -\int_S \nabla u \cdot \nabla v \, \mathrm{d}x \qquad \text{for all } v \in \mathcal{D}(S),$$

cp. p. 116 in [53]. Then, according to Lemma 4.3 in [53], there exists a $g_u \in H^{-1/2}(\partial S)$ such that

$$\int_S f v \, \mathrm{d}x = \int_{\partial S} g_u \, \gamma_D v \, \mathrm{d}s - \int_S \nabla u \cdot \nabla v \, \mathrm{d}x \qquad \text{for all } v \in H^1(S), \tag{2.21}$$

where the first and the second term should be understood as dual pairings. We note that this equality is a generalization of the first Green's identity (2.18). Therefore, a bounded linear extension $\gamma_N : H^1(S) \to H^{-1/2}(\partial S)$ of γ_N from Proposition 2.6 is defined by $\gamma_N : u \mapsto g_u$ with g_u from (2.21). However, the functional g_u depends on f, see the explanation on p. 117 in [53]. We are especially interested in the case that $\Delta u \in L^2(S)$ for some $u \in H^1(S)$. Hence, we define the space

$$H^1_\Delta(S) = \{u \in H^1(S) : \Delta u \in L^2(S)\}. \tag{2.22}$$

Based on (2.21) with $f = \Delta u$ in S and $f = 0$ in $\mathbb{R}^3 \backslash \overline{S}$, we then define by $\gamma_N : u \mapsto g_u$ a unique bounded linear extension $\gamma_N : H^1_\Delta(S) \to H^{-1/2}(\partial S)$. For more details, we refer to Section 2.7 in [62].

Regarding the identity (2.20), the condition $u \in H^1(S, \mathbb{C}^3)$ appears restrictive, since with $v \in H^1(S, \mathbb{C}^3)$ the left-hand side of (2.20) is well-defined for $u \in H(\mathrm{curl}, S)$ and the right-hand side of (2.20) requires simply that $\gamma_t u$ formally defines a bounded functional on $H^{1/2}(\partial S, \mathbb{C}^3)$. So, by enforcing this identity (2.20) for all $u \in H(\mathrm{curl}, S)$ and all $v \in H^1(S, \mathbb{C}^3)$, we define a unique bounded linear extension of γ_t to an operator $\gamma_t : H(\mathrm{curl}, S) \to H^{-1/2}(\partial S, \mathbb{C}^3)$, cf. [13]. Moreover, since the left-hand side of (2.20) remains well-defined likewise for $v \in H(\mathrm{curl}, S)$, one might think of an even more refined identity. In fact, one can prove

Theorem 2.9. *Let $S \subset \mathbb{R}^3$ be a bounded Lipschitz domain and let $Y(\partial S)$ be the space*

$$Y(\partial S) = \{f \in H^{-1/2}(\partial S, \mathbb{C}^3) : \exists u \in H(\mathrm{curl}, S) \text{ with } \gamma_t u = f\}, \tag{2.23}$$

normed by

$$\|f\|_{Y(\partial S)} = \inf_{u \in H(\mathrm{curl}, S), \gamma_t u = f} \|u\|_{H(\mathrm{curl}, S)}.$$

With this norm, $Y(\partial S)$ is a Banach space.

(i) The operator $\gamma_t : H(\mathrm{curl}, S) \to Y(\partial S)$ is surjective and bounded.

(ii) For all $u \in H(\mathrm{curl}, S)$ and $v \in H^1(S, \mathbb{C}^3)$ there holds

$$\int_S (\mathrm{curl}\, u \cdot v - u \cdot \mathrm{curl}\, v)\, \mathrm{d}x = \int_{\partial S} \gamma_t u \cdot \gamma_D v \, \mathrm{d}s. \tag{2.24}$$

(iii) There is a unique bounded extension of γ_T to an operator $\gamma_T : H(\mathrm{curl}, S) \to Y(\partial S)'$ such that for all $u, v \in H(\mathrm{curl}, S)$ there holds

$$\int_S (\mathrm{curl}\, u \cdot v - u \cdot \mathrm{curl}\, v)\, \mathrm{d}x = \int_{\partial S} \gamma_t u \cdot \gamma_T v \, \mathrm{d}s. \tag{2.25}$$

Proof.

(i) This is clear by the definition of $Y(\partial S)$ and its norm. In fact,

$$\|\gamma_t\|_{H(\mathrm{curl},S)\to Y(\partial S)} \leq 1.$$

(ii) The operator γ_t is defined on $H(\mathrm{curl},S)$ such that the identity (2.24) holds.

(iii) Although its statement differs slightly from the one here, careful inspection of the proof of Theorem 3.31 in [55] reveals the assertion. □

The space $Y(\partial S)$ is a proper subset of $H^{-1/2}(\partial S,\mathbb{C}^3)$, cp. Remark 3.30 in [55]. For a detailed characterization of γ_t and its range $Y(\partial S)$, see Theorem 4.1 in [13]. For a Lipschitz polyhedron S, we also mention the earlier article [12]. In addition to the above generalized identities, one has the following result for the normal trace operator.

Theorem 2.10. *Let $S \subset \mathbb{R}^3$ be a bounded Lipschitz domain. The normal trace operator γ_n from Definition 2.7 can be extended by continuity to a bounded linear operator $\gamma_n : H(\mathrm{div},S) \to H^{-1/2}(S)$, and for all $u \in H(\mathrm{div},S)$ and $v \in H^1(S)$ there holds the Green's identity*

$$\int_S (u \cdot \mathrm{grad}\, v + (\mathrm{div}\, u)\, v)\,\mathrm{d}x = \int_{\partial S} \gamma_n u\, \gamma_D v\,\mathrm{d}s. \tag{2.26}$$

A proof is given in [55], see Theorem 3.24 therein. As a reference, we collect the results of this subsection in the following corollary.

Corollary 2.11. *Let $S \subset \mathbb{R}^3$ be a bounded Lipschitz domain. The trace operators are well-defined as bounded linear mappings*

$$\begin{aligned}
\gamma_D &: H^1(S) \to H^{1/2}(\partial S),\\
\gamma_n &: H(\mathrm{div},S) \to H^{-1/2}(S),\\
\gamma_N &: H^2(S) \to H^{1/2}(\partial S),\\
\gamma_N &: H^1_\Delta(S) \to H^{-1/2}(\partial S),\\
\gamma_t &: H(\mathrm{curl},S) \to Y(\partial S),\\
\gamma_T &: H(\mathrm{curl},S) \to Y(\partial S)',
\end{aligned}$$

with $Y(\partial S)$ as defined in (2.23).

Further extensions of the trace operator γ_D from Definition 2.5 are stated in Theorems 3.37 and 3.38 in [53]. We have shown the unique extensions of the trace operators from the classical spaces $C^\infty(\overline{S})$ and $C^\infty(\overline{S},\mathbb{C}^3)$, respectively, to Sobolev spaces on a *bounded* Lipschitz *domain* S. However, in the following we let these operators frequently refer to a Lipschitz set $S \subset \mathbb{R}^3$, i.e. to a possibly unbounded and disconnected S. If S is unbounded, the application of some trace operator is guaranteed to be well-defined by a truncation procedure in the given context. In addition, if S is disconnected, the respective trace operator is meant to denote the corresponding unique trace operator for the (bounded) connected component for which it is evaluated. We will use the notations $u|_{\partial S}$, $\nu \cdot u|_{\partial S}$, $\partial u/\partial \nu$, $\nu \times u|_{\partial S}$, and $(\nu \times u|_{\partial S}) \times \nu$ synonymously with $\gamma_D u$, $\gamma_n u$, $\gamma_N u$, $\gamma_t u$, and $\gamma_T u$, respectively. A particular notation is chosen in any occurrence only for a better readability.

2.5 Additional function spaces

Besides the space $H^s_\alpha(S)$, defined for a Lipschitz set $S \subseteq \Pi$ and $s > 1/2$ in (2.14), we introduce the spaces

$$H_\alpha(\mathrm{div},S) = \overline{\mathcal{E}_\alpha(\overline{S})}^{\|\cdot\|_{H(\mathrm{div},S)}} \qquad \text{and} \qquad H_\alpha(\mathrm{curl},S) = \overline{\mathcal{E}_\alpha(\overline{S})}^{\|\cdot\|_{H(\mathrm{curl},S)}} \tag{2.27}$$

for a Lipschitz set $S \subseteq \Pi$. For a function u in one of the spaces $H^s_\alpha(S)$, $H_\alpha(\mathrm{div},S)$, and $H_\alpha(\mathrm{curl},S)$, we write u_α instead of u. Finally, we define the spaces $H^s_{\alpha,\mathrm{loc}}(S)$, $H_{\alpha,\mathrm{loc}}(\mathrm{div},S)$, and $H_{\alpha,\mathrm{loc}}(\mathrm{curl},S)$. In view of (2.9), one might define $H^s_{\alpha,\mathrm{loc}}(S)$ for a Lipschitz set $S \subseteq \Pi$ and $s > 1/2$ as the set of functions u_α such that $u_\alpha|_{\widetilde{S}} \in H^s_\alpha(\widetilde{S})$ for every compact subset $\widetilde{S}$ of S. The other both spaces could be defined analogously. Now, let $S \subseteq \Pi$ be a Lipschitz set such that $\partial S \cap \partial \Pi \neq \emptyset$ and $S_c = \overline{S_{\mathrm{per}}} \cap \overline{\Pi}$ is not degenerate, hence $(S_c \cap \partial \Pi) \cup \partial S = \partial S$. If the $H_{\alpha,\mathrm{loc}}$-spaces are defined in the above manner then they do not provide a statement about the behavior of a space element at the boundary ∂S. However, we will work with fields on S whose α-quasi-periodic extensions are H^s_α-, $H_\alpha(\mathrm{div})$-, or $H_\alpha(\mathrm{curl})$-regular even across the (partial) boundary $\partial S \cap \partial \Pi$. To capture this feature in the spaces, we define them by

$$\begin{aligned}
H^s_{\alpha,\mathrm{loc}}(S) &= \left\{u_\alpha : S \to \mathbb{C} : U_\alpha\big|_{\widetilde{S}} \in H^s_\alpha(\widetilde{S}) \text{ for any compact } \widetilde{S} \subset S_{\mathrm{per}}\right\},\\
H_{\alpha,\mathrm{loc}}(\mathrm{div},S) &= \left\{u_\alpha : S \to \mathbb{C}^3 : U_\alpha\big|_{\widetilde{S}} \in H_\alpha(\mathrm{div},\widetilde{S}) \text{ for any compact } \widetilde{S} \subset S_{\mathrm{per}}\right\},\\
H_{\alpha,\mathrm{loc}}(\mathrm{curl},S) &= \left\{u_\alpha : S \to \mathbb{C}^3 : U_\alpha\big|_{\widetilde{S}} \in H_\alpha(\mathrm{curl},\widetilde{S}) \text{ for any compact } \widetilde{S} \subset S_{\mathrm{per}}\right\},
\end{aligned}$$

where U_α denotes the α-quasi-periodic extension of u_α, S_{per} is defined according to (2.15), and $s > 1/2$. A few times, we will also need the space

$$H^s_{\alpha,\text{loc}}(S) = \left\{u_\alpha : S \to \mathbb{C} : U_\alpha\big|_{\widetilde{S}} \in H^s_\alpha(\widetilde{S}) \text{ for any compact } \widetilde{S} \subset S^{(2)}_{\text{per}}\right\}$$

and the corresponding counterparts of $H_{\alpha,\text{loc}}(\text{div}, S)$ and $H_{\alpha,\text{loc}}(\text{curl}, S)$ for Lipschitz sets S with $S \subseteq 2 \cdot \Pi$, but $S \not\subseteq \Pi$. The set $S^{(2)}_{\text{per}}$ is given by

$$S^{(2)}_{\text{per}} = \overline{\{x \in \mathbb{R}^3 : x + (2\Lambda) \odot z \in S \text{ for some } z \in \mathbb{Z}^3\}}^{\circ},$$

and the spaces $\mathcal{E}_\alpha(\overline{\widetilde{S}})$, $H^s_\alpha(\widetilde{S})$, $H_\alpha(\text{div}, \widetilde{S})$, and $H_\alpha(\text{curl}, \widetilde{S})$ are defined exactly as in (2.13), (2.14), and (2.27), with $\widetilde{S}$ replacing S and the α-quasi-periodicity still referring to the quasi-period $\Lambda = (2\pi, 2\pi, 0)^T$. Finally, we let

$$H^s_{\alpha,\text{loc}}(\mathbb{R}^3) = \left\{u_\alpha : \mathbb{R}^3 \to \mathbb{C} \ \alpha\text{-q.-p.} : u_\alpha\big|_{\widetilde{S}} \in H^s(\widetilde{S}) \text{ for any compact } \widetilde{S} \subset \mathbb{R}^3\right\},$$

and $H_{\alpha,\text{loc}}(\text{div}, \mathbb{R}^3)$, $H_{\alpha,\text{loc}}(\text{curl}, \mathbb{R}^3)$ similarly.

Chapter 3

The acoustic case

3.1 The direct problem

3.1.1 Problem formulation

We start with a schematic description of the direct problem in the acoustic case. At this point, we do not yet care about the regularity of the contrast $q = n - 1$ and the scattered acoustic field u^s_α. We assume first that the incident acoustic field u^i_α satisfies the Helmholtz equation $\Delta u^i_\alpha + k_0^2 u^i_\alpha = 0$ in the whole unit cell Π. This holds true for a plane wave incidence (see Section 2.1). We now consider the following problem: Given an α-quasi-periodic incident field u^i_α, determine an α-quasi-periodic scattered field u^s_α from the equations

$$\Delta u_\alpha + k_0^2 (1+q)\, u_\alpha = 0 \qquad \text{in } \Pi, \tag{3.1}$$

$$u_\alpha = u^i_\alpha + u^s_\alpha \qquad \text{in } \Pi, \tag{3.2}$$

$$[u_\alpha]_\Gamma = 0, \tag{3.3}$$

$$\left[\frac{\partial u_\alpha}{\partial \nu}\right]_\Gamma = 0, \tag{3.4}$$

and the representation

$$u^s_\alpha(x) = \sum_{z\in Z} u^\pm_z\, e^{\mathrm{i}(\alpha_z\cdot x \pm \beta_z x_3)} \qquad \text{in } R_\pm \tag{3.5}$$

where $Z = \mathbb{Z}^2 \times \{0\}$, $\alpha_z = \alpha + z$, and $\beta_z = \sqrt{k_0^2 - |\alpha_z|^2}$. Later on, we will need that none of the coefficients β_z, $z \in Z$, vanishes. Since in acoustics the wave number k_0 and the frequency ω of the incident field are related by $k_0 = \omega / c_0$, this condition amounts to excluding the frequencies in the set

$$\mathcal{E} = \{\omega \in \mathbb{R}^+ : \omega = c_0\,|\alpha_z| \text{ for some } z \in Z\}, \tag{3.6}$$

the so-called *Rayleigh frequencies*. Obviously, $\mathcal{E}$ is a discrete set and $\omega_z = c_0 |\alpha_z|$ tends to infinity as $|z|$ goes to infinity. The equations (3.3) and (3.4) represent transmission conditions at $\Gamma = \partial\Omega \cap \Pi$ (see (1.1)). Further, ν is the exterior unit normal vector to Ω and $[f]_\Gamma$ denotes the jump $[f]_\Gamma = f|_+ - f|_-$, where $f|_\pm$ is the trace of f on Γ when approaching Γ from the outside and the inside of Ω, respectively. The series in (3.5) is required to converge uniformly on compact subsets of $R_+ \cup R_-$. Here and below, the complex square root is defined as the unique holomorphic extension of the square root on $\mathbb{R}_0^+$ to all of $\mathbb{C}\backslash(-\mathrm{i}\infty, 0)$, i.e. to the complex plane slit at the negative imaginary axis. The representation (3.5) is easily recovered by a Fourier expansion of the periodic field $\widetilde{u}^s(x) = e^{-\mathrm{i}\alpha\cdot x} u_\alpha^s(x)$ and ensuring radiating behavior as well as boundedness of u_α^s. Hence, it acts as a radiation condition on u_α^s and is called the *Rayleigh (expansion) radiation condition*. For given phase shift α and wave number k_0, the so-called *Rayleigh coefficients* $u_z^\pm \in \mathbb{C}$, $z \in Z$, completely determine the field u_α^s. A function which fulfills (3.5) is said to be *radiating*. Together, the equations (3.1)–(3.5) make up the mathematical model for an α-quasi-periodic acoustic transmission problem, with transmission in Ω'. Entailing a first modification of the above problem, we consider a special type of incident fields, which is chosen in a related setting in [44]. Let $\Gamma_i = \Gamma_{i,+} \cup \Gamma_{i,-}$ be an incidence surface where $\Gamma_{i,+} \subset \overline{R_+} \cap \Pi$ and $\Gamma_{i,-} \subset \overline{R_-} \cap \Pi$ are flat surfaces with non-empty relative interiors in the planes containing $\Gamma_{i,+}$ and $\Gamma_{i,-}$, respectively. We assume α-quasi-periodic incident fields which are superpositions of fields generated by acoustic point sources located on Γ_i. Such a field is a classical solution to the Helmholtz equation $\Delta u_\alpha^i + k_0^2 u_\alpha^i = 0$ in the restricted domain $\Pi\backslash\Gamma_i$. For the scattered field u_α^s, we obtain the equation

$$\Delta u_\alpha^s + k_0^2 (1+q) u_\alpha^s = -k_0^2 q u_\alpha^i \qquad \text{in } \Pi. \tag{3.7}$$

By (3.2) and the smoothness of u_α^i in a neighborhood of Γ, the transmission conditions (3.3) and (3.4) imply

$$[u_\alpha^s]_\Gamma = 0 \qquad \text{and} \qquad \left[\frac{\partial u_\alpha^s}{\partial \nu}\right]_\Gamma = 0. \tag{3.8}$$

In the following, we choose $L^\infty(\Pi)$ as the source space for the contrast q. However, to ensure a proper problem formulation and treatment, we actually have to restrict to a modest subset of $L^\infty(\Pi)$. In fact, we require that the boundary of the essential support of q is a Lebesgue null set in $\mathbb{R}^3$, that the periodic extension of q into $\mathbb{R}^3$ has a piecewise smooth representative, and that the (countably many) interfaces between smooth parts of one such representative enclose Lipschitz domains. This class contains virtually all physically relevant contrasts. At all of the interfaces between smooth parts we have to impose transmission conditions of the type (3.8). But, not to put large effort in similar and frequently arising details, we keep this issue in mind and leave the problem formulation at stating the conditions at $\Gamma = \partial\Omega \cap \Pi$. Clearly, the differential equation (3.7) has no classical solution in all of Π in general, and we have to make precise the notion of solutions we are interested in. Moreover, making considerations analog to those above for q, we replace u^i_α on the right-hand side of (3.7) by a 'source' f and let f be an element of $L^2(\Pi)$ which is essentially supported in $\overline{\Omega}$. In this more abstract setting, we write v_α instead of u^s_α.

Precisely, we treat the following generalized **direct problem**: Given $f \in L^2(\Pi)$ with support in $\overline{\Omega}$, find a radiating function v_α which satisfies

$$\Delta v_\alpha + k_0^2(1+q)\,v_\alpha = -k_0^2 q f \qquad \text{in } \Pi \tag{3.9}$$

together with the transmission conditions

$$[\gamma_D v_\alpha]_\Gamma = 0 \qquad \text{and} \qquad [\gamma_N v_\alpha]_\Gamma = 0. \tag{3.10}$$

Here, $[\gamma_D v_\alpha]_\Gamma = \gamma_{D,+} v_\alpha|_\Gamma - \gamma_{D,-} v_\alpha|_\Gamma$, where $\gamma_{D,+}$ and $\gamma_{D,-}$ denote the trace operators for Ω^{ext} and Ω, respectively. The jump of the Neumann trace on Γ is given by $[\gamma_N v_\alpha]_\Gamma = -\gamma_{N,+} v_\alpha|_\Gamma - \gamma_{N,-} v_\alpha|_\Gamma$, where $\gamma_{N,+}$ and $\gamma_{N,-}$ are the Neumann trace operators for Ω^{ext} and Ω, respectively. The minus sign in front of $\gamma_{N,+}$ is due to the fact that $\gamma_{N,+}$ is the generalization of the normal derivative on $\partial\Omega^{\text{ext}}$, where the normal vector points *into* rather than out of Ω on Γ (check against (3.8)).

Variational formulation We understand (3.9) with (3.10) in the variational sense. In fact, for f as given above, we seek a radiating function $v_\alpha \in H^1_{\alpha,\text{loc}}(\Pi)$ such that

$$\int_\Pi \left(\nabla v_\alpha \cdot \nabla \psi_{-\alpha} - k_0^2(1+q)\,v_\alpha\,\psi_{-\alpha}\right) \mathrm{d}x = k_0^2 \int_\Omega q f\,\psi_{-\alpha}\,\mathrm{d}x \tag{3.11}$$

holds for all $\psi_{-\alpha} \in H^1_{-\alpha}(\Pi)$ with compact support with respect to x_3. This latter requirement is meant in the sense that there is a compact set $M \subset \overline{\Pi}$ such that $\psi_{-\alpha}((x_1,x_2,\cdot))$ is (essentially) supported in M for all $(x_1,x_2) \in (-\pi,\pi)^2$. We recall that the α-quasi-periodic extension of a function in $H^1_{\alpha,\text{loc}}(\Pi)$ is H^1_α-regular across the boundary $\partial\Pi$. Now, we make some remarks about the formulation (3.11). All terms in (3.11) are clearly well-defined and v_α naturally satisfies the first transmission condition in (3.10) by the ansatz space $H^1_{\alpha,\text{loc}}(\Pi)$. However, only if a radiating solution v_α to (3.11) is sufficiently regular, this v_α can be interpreted as a variational solution to the problem (3.9) with (3.10). In this case, the jump $[\gamma_N v_\alpha]_\Gamma$ is well-defined (cf. Corollary 2.11) and the transmission condition $[\gamma_N v_\alpha]_\Gamma = 0$ is already accounted for in the formulation (3.11). The latter is seen by separate application of the Green's identity (2.21) to the equation (3.9) in the domains Ω and Ω^{ext} to derive (3.11). In the arising boundary integrals, the contributions on $\partial\Pi$ cancel out. We keep in mind the problem (3.9), (3.10) as the motivation for (3.11).

Closing this subsection, we comment on the existence, uniqueness, and regularity of a variational solution v_α to the direct problem. Let us start with the regularity. An adaptation of a standard interior elliptic regularity result (e.g. [27, Theorem 8.8]) to our problem implies that a weak solution to (3.9), in particular v_α, lies in fact in $H^2_{\alpha,\text{loc}}(\Pi)$. It is even a classical solution in Ω^{ext} ([27, Corollary 8.11] or [69, Weyl's Lemma]) and so analytic there ([16, Theorem 3.5]). This also makes the Rayleigh expansion (3.5) well-defined for v_α. Since Π meets the cone condition, v_α is continuous in Π according to Sobolev's Lemma ([69, Lemma 13.XI]). We confine ourselves here to describing now a common procedure to show existence and uniqueness, rather than conducting a complete proof for our setting. First, one incorporates the Rayleigh radiation condition into a variational formulation by considering the scattering problem in the truncated domain $\Pi\backslash\overline{R_+ \cup R_-}$, instead of in Π as for (3.11), and imposing transparent boundary conditions on the artificial boundaries Γ_+ and Γ_-. These conditions are set via the *Dirichlet-to-Neumann operators* on Γ_+ and Γ_-, obtained by simple formal derivation of the Rayleigh expansion, noting that the unit normal on $\Gamma_\pm$ equals $\pm e_3$, respectively. Here, we have used that the Rayleigh expansion in fact still holds in some neighborhood of $\Gamma_\pm$ (in Π), as it is easily seen by the definition of the sets $R_\pm$. From the fact that $e^{\mathrm{i}\alpha_z\cdot x}$, $z \in Z$, as functions of $(x_1,x_2) \in (-\pi,\pi)^2$ are orthogonal to each other it follows that the Rayleigh coefficients of a solution $\widetilde{v}_\alpha$ to the just described variational problem are uniquely determined. By this, $\widetilde{v}_\alpha$ characterizes completely the behavior in $R_+ \cup R_-$ of any radiating solution v_α to (3.11) which coincides with $\widetilde{v}_\alpha$ in $\Pi\backslash\overline{R_+ \cup R_-}$. One finds that the problem on the truncated domain is equivalent to the formulation (3.11) combined with the radiation condition. Usually, the variational problem is stated for the total field u_α rather than for the scattered field $v_\alpha = u^s_\alpha$. Then, the main step is to prove that the operator induced by the associated sesquilinear form is Fredholm with index zero (on a properly chosen space) and depends holomorphically on the wave number in a set $\mathcal{S} \subset \mathbb{C}$ which we specify in a moment. For the theoretical analysis, the wave number is allowed here to be complex-valued, we therefore write k instead of k_0.

We recall that for a plane wave $u_\alpha^i(x) = p\, e^{\mathrm{i}k\theta\cdot x}$ the phase shift $\alpha \in \mathbb{C}^2 \times \{0\}$ and the wave number k are related by

$$\alpha = k\theta = k(\sin\phi_1 \cos\phi_2, \sin\phi_1 \sin\phi_2, 0)^T, \tag{3.12}$$

where ϕ_1 and ϕ_2 are the angles of incidence (see p. 11). Moreover, the wave number and the frequency ω are related in acoustics by $k = \omega/c$, where c stands for the speed of sound in a homogeneous background medium. One may equivalently assume two of the quantities k, ω, and c to be complex-valued. The set $\mathcal{S} = \mathcal{S}_{(\phi_1,\phi_2)}$ of admissible wave numbers is given by

$$\mathcal{S} = \{k \in \mathbb{C} : k^2 - \alpha_z \cdot \alpha_z \notin (-\mathrm{i}\infty, 0] \text{ for all } z \in Z\}$$

where $\alpha_z = \alpha + z$ with $\alpha = \alpha(k)$ from (3.12). To these wave numbers there obviously correspond the frequencies in $\widetilde{\mathcal{S}} = \{\omega \in \mathbb{C} : \omega/c \in \mathcal{S}\}$. Aside, we remark that the set $\mathcal{E}$ of the Rayleigh frequencies, considered in (3.6) for real-valued wave numbers $k = k_0$ for $c = c_0 \in \mathbb{R}^+$, is given as $\mathcal{E} = \mathbb{R}^+ \backslash \widetilde{\mathcal{S}}$. In addition to the main step, one can show that for certain contrasts q as well as for sufficiently small wave numbers $k_0 > 0$, the direct problem is uniquely solvable (see [65]). Combining the above results on the variational form with such a particular uniqueness result and applying analytic Fredholm theory (cf. [29]) finally yields that the direct problem is uniquely solvable for all $k_0 \in \mathcal{S}^+ \backslash D$ where $\mathcal{S}^+ = \mathcal{S} \cap \mathbb{R}^+$ and D is a discrete subset of $\mathcal{S}^+$ without finite accumulation point. This approach is chosen e.g. in [65] for a smooth contrast and a plane wave source $f = u_\alpha^i$. We only remark here that the arguments therein can, with some modifications, be carried over to our setting.

3.1.2 The Green's function and a representation theorem

For the treatment of the problems in this chapter, we choose an integral equation approach. In a different, yet related setting, this has been suggested in [44]. A central ingredient for this approach is the α-quasi-periodic scalar Green's function for the Helmholtz operator in free field conditions, which can be represented as

$$G_\alpha(y,x) = \frac{\mathrm{i}}{8\pi^2} \sum_{z\in Z} \frac{1}{\beta_z} e^{\mathrm{i}(\alpha_z\cdot(x-y)+\beta_z|x_3-y_3|)} \tag{3.13}$$

for $x, y \in \Pi$ with $x_3 \neq y_3$. At this point our initial assumption enters that $\beta_z \neq 0$ for all $z \in Z$. The restriction $x_3 \neq y_3$ for (3.13) is necessary in order to guarantee absolute convergence of the series. Anyway, the Green's function G_α is well-defined for all $x, y \in \Pi$ with $x \neq y$, and there are more advanced representations of G_α without the limitation $x_3 \neq y_3$, see e.g. [3] and, for the 2D case, the comprehensive article [51]. We use the series form for simplicity. For fixed $y \in \mathbb{R}^3$, $G_\alpha(y,\cdot)$ as a distribution can be interpreted as an array of point sources. In fact, a technical, but straightforward proof reveals that formally there holds

$$\Delta_x G_\alpha(y,x) + k_0^2 G_\alpha(y,x) = -\sum_{z\in Z} e^{2\pi \mathrm{i}\,\alpha\cdot z}\delta_{y+2\pi z}(x). \tag{3.14}$$

This implies that inside the unit cell Π it is

$$\Delta_x G_\alpha(y,x) + k_0^2 G_\alpha(y,x) = -\delta_y(x), \qquad x \in \Pi, \tag{3.15}$$

and analogously, for fixed x,

$$\Delta_y G_\alpha(y,x) + k_0^2 G_\alpha(y,x) = -\delta_x(y), \qquad y \in \Pi. \tag{3.16}$$

It is easy to show that in $(\Pi \times \Pi)\backslash\{(x,x) : x \in \Pi\}$ the function G_α has the form

$$G_\alpha(y,x) = \Phi(x,y) + \Psi(x-y), \tag{3.17}$$

where Φ denotes the fundamental solution to the scalar Helmholtz equation in $\mathbb{R}^3$ and Ψ is a classical solution to the Helmholtz equation in $2\cdot\Pi$ (and hence analytic). Moreover, it follows from the representation (3.13) that G_α satisfies $G_\alpha(y,x) = G_{-\alpha}(x,y)$ and obeys the radiation conditions

$$G_\alpha(y,x) = \sum_{z\in Z} g_z^{\pm}(y)\, e^{\mathrm{i}(\alpha_z\cdot x \pm \beta_z x_3)}, \qquad x_3 \gtrless y_3, \qquad \text{and} \tag{3.18}$$

$$G_\alpha(y,x) = \sum_{z\in Z} g_z^{\pm}(x)\, e^{-\mathrm{i}(\alpha_z\cdot y \pm \beta_z y_3)}, \qquad x_3 \gtrless y_3, \tag{3.19}$$

for fixed y and fixed x, with Rayleigh coefficients $g_z^{\pm}(y)$ and $g_z^{\pm}(x)$, respectively.

Representation theorem

The α-quasi-periodic Green's function is especially important for the following representations of α-quasi-periodic functions in the interior and the exterior of some Lipschitz set.

Theorem 3.1. *Let $L \subset \Pi$ be a bounded Lipschitz set such that L_{per} is also Lipschitz and $L_c = \overline{L_{per}} \cap \overline{\Pi}$ is not degenerate, according to Definition 2.1.*

(i) Assume $v_\alpha \in H^1_{\alpha,\Delta}(L)$. Then there holds Green's formula

$$\int_{\partial L}\left(v_\alpha(y)\frac{\partial}{\partial \nu_y}G_\alpha(y,x) - \frac{\partial v_\alpha}{\partial \nu}(y)\,G_\alpha(y,x)\right)\mathrm{d}s(y) + \\ + \int_L \left(\Delta v_\alpha(y) + k_0^2 v_\alpha(y)\right) G_\alpha(y,x)\,\mathrm{d}y = \begin{cases} -v_\alpha(x), & x \in L \\ 0, & x \in \Pi\backslash\overline{L} \end{cases} \tag{3.20}$$

for almost all $x \in \Pi$. If $v_\alpha \in C^2_\alpha(\overline{L})$, then (3.20) holds for all $x \in \Pi$.

(ii) Assume $v_\alpha \in H^1_{\alpha,loc}(\Pi \backslash \overline{L})$ is a radiating solution to the Helmholtz equation $\Delta v_\alpha + k_0^2 v_\alpha = 0$ in $\Pi \backslash \overline{L}$. Then v_α satisfies

$$\int_{\partial L} \left(v_\alpha(y) \frac{\partial}{\partial \nu_y} G_\alpha(y,x) - \frac{\partial v_\alpha}{\partial \nu}(y)\, G_\alpha(y,x) \right) \mathrm{d}s(y) = \begin{cases} 0, & x \in L \\ v_\alpha(x), & x \in \Pi \backslash \overline{L} \end{cases}. \tag{3.21}$$

Proof (scheme). The proof relies on the arguments in Section 3.3 in [3]. By a sophisticated analysis of the Green's function $G_\alpha(\cdot,x)$, in particular for real wave numbers (like k_0), it is shown there that results from standard potential theory (see e.g. Section 2.2 in [17]) can be transferred to the α-quasi-periodic case. These also yield the identities (3.20) and (3.21). To show the second claim in (i), we modify the proof of Theorem 2.1 in [17] using the second Green's identity for a bounded Lipschitz set, cp. Corollary 3.20 (3) in [55]. In the proof of the identity (3.21) for $x \in L$, the radiation conditions (3.5) for v_α and (3.19) for $G_\alpha(\cdot,x)$ are applied. □

3.1.3 The near field operator

In this subsection, we introduce the so-called *near field operator*, which will be the central object in our following discussion. From the form of this operator it becomes clear that its computation is essentially equivalent to the solution of the direct problem in Ω^{ext}. In a more abstract context, the operator which reflects the direct problem (in Ω^{ext}) is called the *forward operator*.

In our scattering problem, the incident field has the form

$$\widetilde{u}^i_\alpha(x) = \int_{\Gamma_i} G_\alpha(y,x)\,\phi(y)\,\mathrm{d}s(y), \qquad x \in \Pi \backslash \Gamma_i, \tag{3.22}$$

with Γ_i as defined on p. 30. Here, $\phi(y)$ describes the 'moment' of the acoustic point source at $y \in \Gamma_i$, which is represented by $G_\alpha(y,\cdot)$. This function ϕ evidently characterizes the incident field $\widetilde{u}^i_\alpha$. Due to the superposition principle, the scattered acoustic field generated by $\widetilde{u}^i_\alpha$ is given by

$$\widetilde{u}^s_\alpha(x) = \int_{\Gamma_i} \widetilde{u}^s_{p,\alpha}(x,y)\,\phi(y)\,\mathrm{d}s(y), \qquad x \in \Pi,$$

where $\widetilde{u}^s_{p,\alpha}(\cdot,y)$ denotes the response to the incidence of a single point source located at $y \in \Gamma_i$. In practice, this field is measured on some surface $\Gamma_s \subset \Omega^{\text{ext}} =$

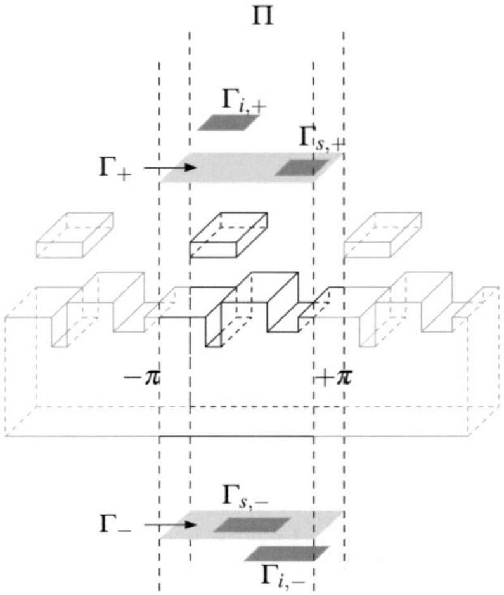

Figure 3.1: Incidence and measurement surfaces

$\Pi\backslash\overline{\Omega}$ not too far away from the medium. This fact is captured in the name 'near field'. Our analysis requires that Γ_s is the union of two flat surfaces $\Gamma_{s,+} \subseteq \Gamma_+$ and $\Gamma_{s,-} \subseteq \Gamma_-$ with non-empty relative interiors in Γ_+ and Γ_-, respectively. Besides, we note that each of the sets $\Gamma_{i,+}$, $\Gamma_{i,-}$, $\Gamma_{s,+}$, and $\Gamma_{s,-}$ might be disconnected. Figure 3.1 exemplifies the geometric situation. The operator which maps the moment function ϕ of the incident field $\widetilde{u}^i_\alpha$ to the scattered field $\widetilde{u}^s_\alpha$ on Γ_s is called the *near field operator* $\widetilde{M} : L^2(\Gamma_i) \to L^2(\Gamma_s)$, reading

$$(\widetilde{M}\phi)(x) = \int_{\Gamma_i} \widetilde{u}^s_{p,\alpha}(x,y)\,\phi(y)\,\mathrm{d}s(y), \qquad x \in \Gamma_s. \tag{3.23}$$

An approximant for the operator $\widetilde{M}$ is computed in practice from the given data ϕ and a discrete set of measurements of $\widetilde{u}^s_\alpha$ on Γ_s. However, in order for the Factorization Method to be applicable later, it is necessary that the near field operator maps a Hilbert space into its dual. This requirement obviously is not satisfied if the incidence surface Γ_i and the measurement surface Γ_s are different. To overcome these problems, in [44] an auxiliary forward operator (a far field operator therein) is used, which is computed approximately from the original, physical one. We adapt this procedure here and define the somewhat artificial near field operator $M : L^2(\Gamma_s) \to L^2(\Gamma_s)$ by

$$(M\varphi)(x) = \int_{\Gamma_s} u^s_{p,\alpha}(x,y)\,\varphi(y)\,\mathrm{ds}(y), \qquad x \in \Gamma_s, \tag{3.24}$$

where $u^s_{p,\alpha}(\cdot,y)$ is the response to the complex conjugate point source $\overline{G_{-\alpha}(y,\cdot)}$ at $y \in \Gamma_s$. This means that $M\varphi$ corresponds to the α-quasi-periodic incident field

$$u^i_\alpha(x) = \int_{\Gamma_s} \overline{G_{-\alpha}(y,x)}\,\varphi(y)\,\mathrm{ds}(y), \qquad x \in \Pi\backslash\Gamma_s. \tag{3.25}$$

The radiation properties of $G_{-\alpha}(y,\cdot) = G_\alpha(\cdot,y)$ (see (3.19)) imply that u^i_α satisfies

$$u^i_\alpha(x) = \sum_{z\in Z} u^\pm_z(\varphi)\,e^{\mathrm{i}(\alpha_z\cdot x \mp \overline{\beta_z} x_3)}, \qquad x_3 \gtrless m_\pm, \tag{3.26}$$

where

$$u^\pm_z(\varphi) = -\frac{\mathrm{i}}{8\pi^2}\frac{1}{\overline{\beta_z}}\int_{\Gamma_s} \varphi(y)\,e^{-\mathrm{i}(\alpha_z\cdot y \mp \overline{\beta_z} y_3)}\,\mathrm{ds}(y).$$

The reason why we have to restrict the representation (3.26) to $R_+ = \{x \in \Pi : x_3 > m_+\}$ and $R_- = \{x \in \Pi : x_3 < m_-\}$ is that Ω^{ext} might be connected. In this case, incident waves originating on $\Gamma_{s,+}$ and $\Gamma_{s,-}$ can propagate to Γ_- and Γ_+, respectively. For $x \in \Pi$ inbetween Γ_+ and Γ_-, neither $x_3 > y_3$ nor $x_3 < y_3$ holds for all $y \in \Gamma_s$. However, we can decompose u^i_α as the sum of $u^i_{\alpha,+}$ and $u^i_{\alpha,-}$ defined by

$$u^i_{\alpha,\pm} = \int_{\Gamma_{s,\pm}} \overline{G_{-\alpha}(y,x)}\,\varphi(y)\,\mathrm{ds}(y),$$

which allow a more detailed representation of u^i_α by

$$u^i_{\alpha,+}(x) = \sum_{z\in Z} u^\pm_{z,+}(\varphi)\,e^{\mathrm{i}(\alpha_z\cdot x \mp \overline{\beta_z} x_3)}, \qquad x_3 \gtrless m_+, \tag{3.27}$$

and

$$u^i_{\alpha,-}(x) = \sum_{z\in Z} u^\pm_{z,-}(\varphi)\,e^{\mathrm{i}(\alpha_z\cdot x \mp \overline{\beta_z} x_3)}, \qquad x_3 \gtrless m_-, \tag{3.28}$$

with evident coefficients $u^\pm_{z,+}(\varphi)$ and $u^\pm_{z,-}(\varphi)$. We only remark that the signs of the x_3-terms in the exponents in (3.27) and (3.28) are flipped compared to those of the corresponding terms in the expressions for the physical incident field $\widetilde{u}^i_\alpha$, defined in (3.22). It means that the waves caused by u^i_α are travelling in the opposite x_3-direction. The approximation of the auxiliary near field operator M, which we will work with, by means of the physical near field operator $\widetilde{M}$ is discussed in detail in Chapter 5.

3.2 The inverse problem

With the above preparations, we now address our main goal, referring to a problem which is 'inverse' to the problem described in Section 3.1. At first go, one might think that it is about the inversion of the operator $\widetilde{M}$, whose computation represents in a way the direct problem, or to a full reconstruction of the contrast, which governs the direct problem. However, the inversion of $\widetilde{M}$ is useless since its argument, the moment function ϕ, is the known input in practice. The full reconstruction of the contrast is a sensible, but also the most ambitious objective. In many applications, the focus is instead on the determination or the optimization of the shape of some object.

This motivates the following **inverse problem**: Given the scattered fields $\widetilde{u}^s_\alpha$ on Γ_s for all moment functions $\phi \in L^2(\Gamma_i)$ (and a single fixed wave number k_0), find the shape of the scattering medium or, equivalently, the support of the contrast q!

To be precise, by the 'shape' of the medium we always mean its actual shape together with its location in $\mathbb{R}^3$. We note that, due to superposition, knowing the scattered fields $\widetilde{u}^s_\alpha$ (on Γ_s) for all $\phi \in L^2(\Gamma_i)$ is equivalent to knowing the responses $\widetilde{u}^s_{p,\alpha}(\cdot, y)$ (on Γ_s) for all $y \in \Gamma_i$. Anticipating the unique solvability of the above inverse problem, one might apply some iterative method to solve it. For inverse scattering from bounded inhomogeneous media in $\mathbb{R}^3$, one can prove that even the contrast q, not only its support, is uniquely determined by the complete far field patterns for all directions of incident plane waves, see e.g. Theorem 4.3 in [45]. A similar rigorous uniqueness result for periodic contrasts seems to be still an open issue. An iterative method for the full reconstruction of q from far field data for a bounded inhomogeneity is proposed in [33]. For the reconstruction of the profile of a perfectly reflecting lamellar grating, an iterative method based on the so-called domain derivative of the scattered field is established in [30]. On the one hand, iterative methods are powerful and widely applicable, on the other hand they have some inconvenient requirements, including conditions on the forward operator, a reasonable initial estimate of the (unique) solution, or so-called source conditions to guarantee convergence rates, cf. [25, 32]. Moreover, on the downside of being highly non-specific, these methods do not utilize directly the characteristics of the problem at hand. For the adequate goal of reconstructing the support of the contrast, there are more suitable (non-iterative) alternatives which in fact do not require an initial estimate, see, e.g., [15] for qualitative methods in inverse scattering theory. A well-known such alternative is the so-called *Factorization Method*. It has been proposed by KIRSCH in [39] for inverse scattering from a bounded obstacle and extended to a class of inverse elliptic problems in [42]. Under modest assumptions, this method allows an efficient and easy-to-implement identification of the support of the respective contrast. It is based on a factorization of the forward operator and a substantial use of the properties of its factors, thereby taking account of the nature of the problem. In the last decade, the Factorization Method was developed further by several people, we refer to the bibliography of the monograph [45]. This book provides a thorough discussion of the method along with different applications. The problem of inverse scattering from homogeneous periodic media is treated in the articles [5, 4] and the thesis [49]. So far, the Factorization Method has not been investigated in the context of inverse scattering from inhomogeneous periodic media. Our intention is to fill this gap.

3.2.1 Factorization of the near field operator

Let us start by briefly sketching the idea of the Factorization Method. The first main feature of the method is the characterization of the range $\mathcal{R}(B^*)$ of the adjoint operator B^* of B from a certain factorization $A = B^* C B$ in terms of the operator A. This relies on a functional analytic result and depends on the properties of the operators B and C. In our discussion, the operator A is the artificial near field operator M and depends implicitly on the medium via the response $u^s_{p,\alpha}$. The second feature of the Factorization Method, specific to the application in inverse scattering problems, is expressed in the proof that for some factorization $M = B^* C B$ there is a deep connection between the range of B^* and the shape of the scattering medium. The combination of both relations then enables us to identify the shape of the medium by means of the artificial near field operator M.

In the following, we make some general assumptions, which we collect for reference in

Assumptions 3.2.

- The sets Ω' and $\Omega = \Omega' \cap \Pi$ are Lipschitz, and the set $\Omega_c = \overline{\Omega_{\text{per}}} \cap \overline{\Pi}$ is not degenerate.
- The connected components of Ω' are simply connected.
- The contrast q is essentially bounded, i.e. $q \in L^\infty(\Pi)$, and satisfies

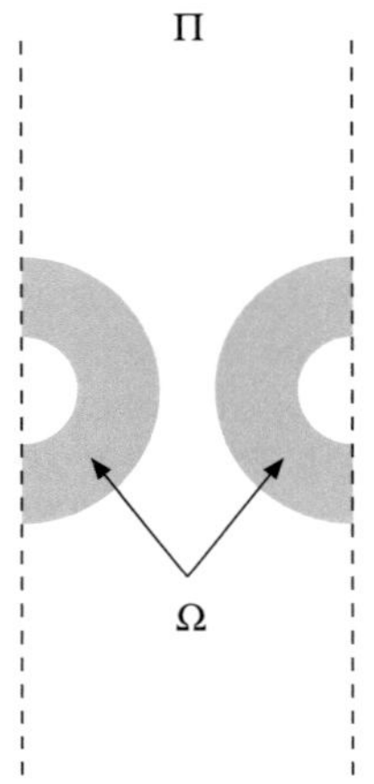

Figure 3.2: the connected components of $\Omega \subset \mathbb{R}^2$ are simply connected

(i) $q = 0$ almost everywhere (a.e.) in Ω^{ext},

(ii) $1 + \operatorname{Re} q = \operatorname{Re} n \geq c_0$ a.e. in Ω for some constant $c_0 > 0$,

(iii) $\operatorname{Im} q \geq 0$ a.e. in Ω,

(iv) $|q|$ is locally bounded from below in Ω, meaning that for every compact subset $S \subset \Omega$ there is a constant $c_S > 0$ such that $|q| \geq c_S$ a.e. in S.

- The direct problem, defined on p. 31, is uniquely solvable.

We remark that, since Ω' is Lipschitz, Ω_{per} coincides with Ω'. Due to the third assumption, $\overline{\Omega}$ does not only contain, but is equal to $\operatorname{supp} q \cap \overline{\Pi}$. Moreover, the second and the third assumption imply that Ω' does not show any inclusions of the background medium (where q vanishes). For this to hold, we have to require the simple connectivity of the connected components of Ω' rather than of those of the set Ω, to exclude cases similar to the 2D example shown in Figure 3.2. Here, the connected components of Ω are simply connected, but those of Ω' (which are rings) are not. Since every Lipschitz set fulfills the cone condition, the embedding of $H^1_\alpha(\Omega)$ into $L^2(\Omega)$ is compact according to the Rellich-Kondrachov theorem, cf. [2]. Now, in the style of the incident field (3.25), we define the integral operator

$H_{\Gamma_s}: L^2(\Gamma_s) \to L^2(\Omega)$ by

$$(H_{\Gamma_s}\varphi)(x) = \sqrt{|q(x)|}\int_{\Gamma_s} \overline{G_{-\alpha}(y,x)}\,\varphi(y)\,\mathrm{d}s(y), \qquad x \in \Omega. \tag{3.29}$$

The meaning of the weighting factor $\sqrt{|q|}$ in (3.29) will be clarified in the proof of Theorem 3.8. The adjoint $H^*_{\Gamma_s}: L^2(\Omega) \to L^2(\Gamma_s)$ of H_{Γ_s} is found to be given by

$$(H^*_{\Gamma_s} g)(x) = \int_\Omega G_{-\alpha}(x,y)\, g(y)\,\sqrt{|q(y)|}\,\mathrm{d}y, \qquad x \in \Gamma_s. \tag{3.30}$$

Moreover, we define the *solution operator* $G: L^2(\Omega) \to L^2(\Gamma_s)$ which maps $\hat{f} \in L^2(\Omega)$ to $v_\alpha|_{\Gamma_s}$ where v_α is radiating and satisfies

$$\begin{aligned} &\int_\Pi \left(\nabla v_\alpha \cdot \nabla \psi_{-\alpha} - k_0^2(1+q)\, v_\alpha\, \psi_{-\alpha}\right)\mathrm{d}x = k_0^2 \int_\Omega \frac{q}{\sqrt{|q|}}\hat{f}\,\psi_{-\alpha}\,\mathrm{d}x \\ \iff\quad &\int_\Pi \left(\nabla v_\alpha \cdot \nabla \psi_{-\alpha} - k_0^2\, v_\alpha\, \psi_{-\alpha}\right)\mathrm{d}x = k_0^2 \int_\Omega \left(\frac{q}{\sqrt{|q|}}\hat{f} + q\, v_\alpha\right)\psi_{-\alpha}\,\mathrm{d}x \end{aligned} \tag{3.31}$$

for all $\psi_{-\alpha} \in H^1_{-\alpha}(\Pi)$ with compact support with respect to x_3. The equation (3.31) resembles the variational formulation (3.11), here $\hat{f}$ plays the role of $\sqrt{|q|}\,f|_\Omega$ in (3.11). According to Assumptions 3.2, there is a unique radiating solution to (3.31), hence the operator G is well-defined. Inspecting the above definitions, one observes that the near field operator M can be written as

$$M = G H_{\Gamma_s}. \tag{3.32}$$

Finally, we define the operator $T: L^2(\Omega) \to L^2(\Omega)$ by

$$T\hat{f} = k_0^2\,\mathrm{sign}(q)\left(\hat{f} + \sqrt{|q|}\,v_\alpha\right)\big|_\Omega. \tag{3.33}$$

Here, $\mathrm{sign}(z) = z/|z|$ denotes the complex sign of $z \in \mathbb{C}$ and $v_\alpha \in H^1_{\alpha,\mathrm{loc}}(\Pi)$ is the variational solution to the direct problem with source f, i.e. the radiating solution to (3.31), where $\hat{f}$ is given as the argument of T and related to the source f as described above. Then, (3.31) is easily seen to be the variational form of

$$\Delta v_\alpha + k_0^2 v_\alpha = -k_0^2 \frac{q}{\sqrt{|q|}}\hat{f} - k_0^2 q\, v_\alpha = -\sqrt{|q|}\,T\hat{f} \tag{3.34}$$

in Π, where the right-hand side is extended by zero into Ω^{ext}. In a next step, we show that (3.31) is equivalent to a formulation as an integral equation. Based on this integral equation, we will refine the factorization (3.32) of M and, along the way, prove the well-known *(acoustic) Lippmann-Schwinger equation*. The main tool is established in the following proposition.

Proposition 3.3. *Let W be the volume potential operator defined by*

$$(Wg)(x) = \int_{\Omega} G_{-\alpha}(x,y)\, g(y)\, \sqrt{|q(y)|}\, \mathrm{d}y \tag{3.35}$$

with $g \in L^2(\Omega)$ and $x \in \Pi$. We consider the potential $w_\alpha = Wg$ for some g.

(i) *Let the density $\sqrt{|q|}\,g$ be bounded and Hölder continuous, $\sqrt{|q|}\,g \in C^{0,\gamma}(\Omega)$ with $0 < \gamma \leq 1$. Then the potential $w_\alpha = Wg$ is in $C^2_\alpha(\Pi\backslash\Gamma) \cap C^1_\alpha(\Pi)$ and a classical radiating solution to*

$$\Delta w_\alpha + k_0^2 w_\alpha = -\sqrt{|q|}\,g \qquad \text{in } \Pi\backslash\Gamma. \tag{3.36}$$

The right-hand side of (3.36) is extended by zero into Ω^{ext}.

(ii) *For densities $\sqrt{|q|}\,g \in L^2(\Omega)$, in particular for $q \in L^\infty(\Omega)$ and $g \in L^2(\Omega)$, the potential w_α is in $H^1_{\alpha,loc}(\Pi)$ and a radiating variational solution to (3.36).*

(iii) *The mapping of g to the restriction of w_α to Ω defines a bounded linear operator from $L^2(\Omega)$ to $H^1_\alpha(\Omega)$.*

Proof.

(i) We recall the decomposition (3.17) of the Green's function G_α and the fact that the fundamental solution $\Phi_{k_0} = \Phi$ to the Helmholtz equation in $\mathbb{R}^3$ is the product of the fundamental solution Φ_0 to the Laplace equation in $\mathbb{R}^3$ and a smooth function. Using this, the regularity of w_α and the equation (3.36) can be shown in a similar way as Lemmata 4.1 and 4.2 in [27]. See also p. 141 (a) in [69] and Theorem 8.1 in [17]. The potential w_α inherits the radiating behavior from the integration kernel of W. Besides, we want to emphasize that (3.36) does not imply that $\sqrt{|q|}\,g$ is (or needs to be assumed) α-quasi-periodic since (3.36) is asserted only to hold in $\Pi\backslash\Gamma$ and we have not yet made any statement about the regularity of w_α across the boundary $\partial\Pi$. If the α-quasi-periodic extension of $\sqrt{|q|}\,g$ is $C^{0,\gamma}$-regular across $\partial\Pi\backslash\partial\Omega'$, then (the α-quasi-periodic extension of) w_α is in $C^2_\alpha(\mathbb{R}^3\backslash\partial\Omega') \cap C^1_\alpha(\mathbb{R}^3)$ and (3.36) holds in $\mathbb{R}^3\backslash\partial\Omega'$.

(ii) Concerning the regularity of w_α, we refer to Theorem 8.2 in [17], again under consideration of the relation (3.17). Since a classical solution is also a variational solution, the second part of the assertion results from the denseness of the set of bounded $C^{0,\gamma}(\Omega)$-functions in $L^2(\Omega)$ with respect to the norm of the latter. We also point here to the regularity discussion in Subsection 3.1.1.

(iii) This is a consequence of the definition of w_α and part (ii). □

Obviously, the operators W from (3.35) and $H^*_{\Gamma_s}$ from (3.30) are related by

$$H^*_{\Gamma_s} g = (Wg)\big|_{\Gamma_s}, \qquad g \in L^2(\Omega).$$

By Proposition 3.3 (ii) and the identity $G_{-\alpha}(x,y) = G_\alpha(y,x)$, a solution to the integral equation

$$\begin{aligned} v_\alpha(x) &= k_0^2 \int_\Omega G_\alpha(y,x) \left(\frac{q(y)}{\sqrt{|q(y)|}} \hat{f}(y) + q(y)\, v_\alpha(y) \right) \mathrm{d}y \\ &= k_0^2 \int_\Omega G_\alpha(y,x)\, q(y)\, (f(y) + v_\alpha(y))\, \mathrm{d}y \end{aligned} \tag{3.37}$$

in Π is a radiating solution to (3.31). The equation (3.37) is called the α-quasi-periodic *(acoustic) Lippmann-Schwinger equation.* Vice versa, since the only radiating solution to $\Delta\widetilde{v}_\alpha + k_0^2 \widetilde{v}_\alpha = 0$ in Π is $\widetilde{v}_\alpha \equiv 0$, the unique solution to (3.31) satisfies (3.37). Hence, (3.31) and (3.37) are equivalent, and we can write (3.37) for short as

$$v_\alpha(x) = \int_\Omega G_\alpha(y,x)\, (T\hat{f})(y)\, \sqrt{|q(y)|}\, \mathrm{d}y. \tag{3.38}$$

From this and (3.30), we realize the identity $H^*_{\Gamma_s} T\hat{f} = v_\alpha|_{\Gamma_s} = G\hat{f}$. Finally, inserting this into (3.32), we obtain a factorization of the artificial near field operator M in the form

$$M = H^*_{\Gamma_s} T\, H_{\Gamma_s}. \tag{3.39}$$

We complete the description of the integral equation approach in a corollary to Proposition 3.3. It just rephrases for the periodic case the comments on p. 92 in [45].

Corollary 3.4. *Let Assumptions 3.2 hold.*

(i) If $v_\alpha \in H^1_{\alpha,loc}(\Pi)$ is a radiating solution to (3.31), then the restriction $v_\alpha|_\Omega \in H^1_\alpha(\Omega)$ solves the equation

$$\widetilde{v}_\alpha = k_0^2 W \left(\frac{q}{\sqrt{|q|}} (f + \widetilde{v}_\alpha) \right) \Bigg|_\Omega . \tag{3.40}$$

(ii) If $v_\alpha \in H^1_\alpha(\Omega)$ solves (3.40), then it can be extended by the right-hand side of (3.37) to a radiating solution to (3.31).

Our final result in this subsection deals with some properties of the operator H_{Γ_s} and its adjoint, which are necessary for the setup of a suitable Factorization Method. A thorough analysis of the inner operator T in (3.39), defined in (3.33), is the subject of Subsection 3.2.3.

Proposition 3.5.

(i) The operators H_{Γ_s} and $H_{\Gamma_s}^*$ are compact.

(ii) The operator H_{Γ_s} is injective.

Proof.

(i) We recall that $H_{\Gamma_s}^*$ is given by

$$(H_{\Gamma_s}^* g)(x) = \int_\Omega G_{-\alpha}(x,y)\, g(y)\, \sqrt{|q(y)|}\, \mathrm{d}y$$

with $x \in \Gamma_s$. This is a Hilbert-Schmidt integral operator with kernel in $L^2(\Gamma_s \times \Omega)$ and thus compact, cf. Theorem 7.83 in [59]. The compactness of H_{Γ_s} is a direct consequence, see e.g. Theorem 4.19 in [60].

(ii) Let $\varphi \in \ker H_{\Gamma_s}$. Since $q \neq 0$ a.e. in Ω, we conclude that the potential $h_\alpha : \Pi \to \mathbb{C}$ defined by

$$h_\alpha(x) = \int_{\Gamma_s} \overline{G_{-\alpha}(y,x)}\, \varphi(y)\, \mathrm{d}s(y), \qquad x \in \Pi,$$

vanishes in Ω, cp. (3.29). An analytic continuation argument then shows that $h_\alpha = 0$ in $\{x \in \Pi : m_- < x_3 < m_+\}$, where m_- and m_+ have been introduced on p. 11. Since for $\varphi \in L^2(\Gamma_s)$ there holds $h_\alpha \in H^1_{\alpha,\mathrm{loc}}(\Pi)$ (cf. Theorem 6.11 in [53]), h_α does not jump across Γ_+ and Γ_-. This yields $\gamma_{D,+} h_\alpha = 0$ on $\Gamma_+ \cup \Gamma_-$, where $\gamma_{D,+}$ denotes the trace operator for $R_+ \cup R_-$. Moreover, h_α solves the Helmholtz equation in $R_+ \cup R_-$ and obeys the expansion (3.26). Given these equations, the problem to determine h_α in $R_+ \cup R_-$ is an unusual, α-quasi-periodic *exterior Dirichlet problem*. Compare the classical exterior Dirichlet problem for a bounded region, treated in [16]. This problem has at most one solution, which is seen by the smoothness of a solution to the Helmholtz equation and the representation (3.26) together with the fact that $e^{\mathrm{i}(\alpha_z \cdot x)}$ as well as $e^{-\mathrm{i}(\alpha_z \cdot x)}$ with $z \in Z$ form a basis of $L^2(\Gamma_r)$ for every $\Gamma_r = \{x \in \Pi : x_3 = r\}$ with $r \in \mathbb{R}$. We conclude that h_α vanishes in $R_+ \cup R_-$. Finally, by the jump relation $[\gamma_N h_\alpha]_{\Gamma_s} = -\varphi$ ([53, Theorem 6.11]) we obtain $\varphi \equiv 0$. Hence, H_{Γ_s} is injective. □

3.2.2 The interior transmission eigenvalue problem

Before we proceed with the analysis of the inner operator T in (3.39), we present here a special kind of an eigenvalue problem called the *interior transmission eigenvalue problem*. This problem has some analogy to a transmission problem, but while in the latter the interior as well as the exterior of some domain are involved, in the former the 'transmission' manifests in a coupling at the boundary of a domain of two functions which are both defined in the interior of the domain. It turns out that this special problem, stated for our setting, affects some properties of the near field operator M. To begin, we introduce the equation system

$$\left.\begin{array}{rlrll} \Delta v_\alpha + k_0^2(1+q)\,v_\alpha &= 0, & \Delta w_\alpha + k_0^2 w_\alpha &= 0 & \text{in } \Omega \\ \gamma_D v_\alpha &= \gamma_D w_\alpha, & \gamma_N v_\alpha &= \gamma_N w_\alpha & \text{on } \Gamma \end{array}\right\}, \tag{3.41}$$

where γ_D is the trace operator for Ω and γ_N is the Neumann trace operator for Ω. In a classical formulation for a sufficiently smooth contrast, k_0^2 is said to be an *interior transmission eigenvalue* if there is a nontrivial solution $(v_\alpha, w_\alpha) \in (C_\alpha^2(\Omega) \cap C_\alpha^1(\overline{\Omega}))^2$ to (3.41). For a contrast $q \in L^\infty(\Pi)$, the problem (3.41) has, of course, to be understood in the variational sense.

Definition 3.6. The value k_0^2 is said to be an *interior transmission eigenvalue* with corresponding *eigenpair* $(v_\alpha, w_\alpha) \in H_\alpha^1(\Omega) \times H_\alpha^1(\Omega)$ if $(v_\alpha, w_\alpha) \neq (0,0)$ and (v_α, w_α) satisfies

$$\int_\Omega \left(\nabla v_\alpha \cdot \nabla \psi_{-\alpha} - k_0^2(1+q)\,v_\alpha\,\psi_{-\alpha}\right)\mathrm{d}x = \int_\Omega \left(\nabla w_\alpha \cdot \nabla \psi_{-\alpha} - k_0^2 w_\alpha\,\psi_{-\alpha}\right)\mathrm{d}x \tag{3.42}$$

for all $\psi_{-\alpha} \in H_{-\alpha}^1(\Omega)$ and

$$\int_\Omega \left(\nabla w_\alpha \cdot \nabla \psi_{-\alpha} - k_0^2 w_\alpha\,\psi_{-\alpha}\right)\mathrm{d}x = 0 \tag{3.43}$$

for all $\psi_{-\alpha} \in H_{-\alpha,0}^1(\Omega)$.

Under sufficient regularity of the functions v_α and w_α of an eigenpair (v_α, w_α) is a variational solution to the system (3.41). The coupling boundary condition $\gamma_N v_\alpha = \gamma_N w_\alpha$ on Γ is then implied by the formulation (3.42), (3.43), due to the claim that (3.42) holds for all $\psi_{-\alpha} \in H_{-\alpha}^1(\Omega)$. This can be seen by applying the first Green's identity (2.18) separately to the first both equations in (3.41) and comparing the resulting equations with (3.42), (3.43). The arguments here are analog to those in the short discussion following the variational equation (3.11) for the direct problem. Now, we shortly address a type of contrast for which there are no interior transmission eigenvalues. Let $U \subseteq \Omega$ be some non-empty open set which has non-empty intersection with each connected component of Ω. If $\operatorname{Im} q > 0$ on U, then there is no interior transmission eigenvalue $k_0^2 > 0$. This can be proven using the fact that any variational solution v_α to the first equation in (3.41) lies in $H_\alpha^2(\Omega_0)$ for any subdomain $\Omega_0 \subset \Omega$ [27, Theorem 8.8] and applying a differential inequality for v_α for unique continuation, cf. Lemma 4.15 in [55]. An important consequence of the absence of such eigenvalues for the near field operator M is stated in the following proposition.

Proposition 3.7. *Assume that k_0^2 is not an interior transmission eigenvalue. Then $M : L^2(\Gamma_s) \to L^2(\Gamma_s)$ is injective and has dense range $\mathcal{R}(M)$ in $L^2(\Gamma_s)$.*

Proof. We use the idea of the proof of Theorem 1.8 (d) in [45]. Let $\varphi \in \ker M$. The radiation condition (3.5), an analytic continuation argument, and the fact that Ω' has no inclusions of the background medium yield that the scattered field u_α^s with the near field $M\varphi \equiv 0$ vanishes in Ω^{ext}. From the definition of the operator M, it is clear that u_α^s is raised by the incident field $u_\alpha^i = \int_{\Gamma_s} \overline{G_{-\alpha}(y,\cdot)}\, \varphi(y)\, \mathrm{ds}(y)$. Now, the functions $v_\alpha = u_\alpha^i + u_\alpha^s$ and $w_\alpha = u_\alpha^i$ satisfy the interior transmission eigenvalue problem (3.41). By the assumption on k_0^2, v_α and w_α have to vanish in Ω. As a consequence of the form of the incident field u_α^i and the injectivity of H_{Γ_s} (see Proposition 3.5 (ii)), this is possible only for $\varphi \equiv 0$. Hence, M is injective. For the second part of the assertion, we use the common identity $\mathcal{R}(M)^\perp = \ker M^*$ and show that also the adjoint M^* is injective. This is easily seen to be given by

$$(M^*\varphi)(x) = \int_{\Gamma_s} \overline{u_{p,\alpha}^s(y,x)}\, \varphi(y)\, \mathrm{ds}(y) = \overline{\int_{\Gamma_s} u_{p,\alpha}^s(y,x)\, \overline{\varphi(y)}\, \mathrm{ds}(y)}, \qquad x \in \Gamma_s.$$

The definition of $u_{p,\alpha}^s$ and the form of the Green's function imply the *reciprocity relation* $u_{p,\alpha}^s(y,x) = u_{p,\alpha}^s(-x,-y)$ for all $x, y \in \Gamma_s$, hence

$$(M^*\varphi)(x) = \overline{\int_{\Gamma_s} u_{p,\alpha}^s(-x,-y)\, \overline{\varphi(y)}\, \mathrm{ds}(y)}, \qquad x \in \Gamma_s. \tag{3.44}$$

Now, let $\varphi \in \ker M^*$. We define M_- like M with $-\Gamma_s$ replacing Γ_s and note that $-\Gamma_s$ as well as Γ_s are contained in $\Gamma_+ \cup \Gamma_-$. With (3.44), we obtain $M_-\varphi_- = 0$ on $-\Gamma_s$ where $\varphi_-(y) = \overline{\varphi(-y)}$. By similar arguments as above, M_- is shown to be injective. Therefore, $\varphi_- \equiv 0$ on $-\Gamma_s$ and, accordingly, $\varphi \equiv 0$ on Γ_s. This completes the proof. □

3.2.3 The inner operator

Complementing the preceding considerations, we address now the inner operator T, which remains to be examined in the factorization (3.39) of the artificial near field operator M. Properties of the inner operator are crucial in the functional analytic foundation of the Factorization Method as presented in [45]. Hence, for an application of the Factorization Method it is a key step to prove that the inner operator in a suitable factorization of the respective forward operator (here M) has the required properties. Our next theorem collects the results for the operator T from (3.39).

Theorem 3.8. *Let Assumptions 3.2 hold.*

(i) The operator T defined in (3.33) can be written in the form $T = \widetilde{T} + K$ where $\widetilde{T} : L^2(\Omega) \to L^2(\Omega)$ is defined by $\widetilde{T}\hat{f} = k_0^2 \operatorname{sign}(q)\hat{f}$ and $K : L^2(\Omega) \to L^2(\Omega)$ is compact. If there exist two constants $t \in [0, 2\pi)$ and $c_0 > 0$ such that $\operatorname{Re}(e^{\mathrm{i}t}q) \geq c_0|q|$ holds a.e. in Ω, then the operator $\operatorname{Re}(e^{\mathrm{i}t}\widetilde{T})$ is coercive, precisely,

$$\operatorname{Re}\left(e^{\mathrm{i}t}\langle \widetilde{T}\hat{f}, \hat{f}\rangle_{L^2}\right) \geq k_0^2 c_0 \, \|\hat{f}\|_{L^2}^2 \tag{3.45}$$

holds for all $\hat{f} \in L^2(\Omega)$.

(ii) The operator $\operatorname{Im} T$ is positive semi-definite, i.e. $\operatorname{Im}\langle T\hat{f}, \hat{f}\rangle_{L^2} \geq 0$ for all $\hat{f} \in L^2(\Omega)$.

(iii) Assume that there is a constant $c_1 > 0$ such that $\operatorname{Im} q \geq c_1|q|$ holds a.e. in Ω. Then $\operatorname{Im} T$ is coercive, meaning that there is a constant $c_2 > 0$ such that

$$\operatorname{Im}\langle T\hat{f}, \hat{f}\rangle_{L^2} \geq c_2 \, \|\hat{f}\|_{L^2}^2 \tag{3.46}$$

applies for all $\hat{f} \in L^2(\Omega)$.

(iv) The operator T is injective.

Proof.

(i) The form $T = \widetilde{T} + K$ is obvious where K maps $\hat{f} \in L^2(\Omega)$ to $k_0^2 q/\sqrt{|q|}\, v_\alpha|_\Omega$ and v_α is the radiating solution to the variational equation (3.31). The operator K is compact due to the compact embedding of $H^1_\alpha(\Omega) \ni v_\alpha|_\Omega$ into $L^2(\Omega)$. Further, we get

$$\operatorname{Re}\left(e^{\mathrm{i}t}\langle \widetilde{T}\hat{f}, \hat{f}\rangle_{L^2}\right) = k_0^2 \int_\Omega \frac{\operatorname{Re}(e^{\mathrm{i}t}q)}{|q|} |\hat{f}|^2 \,\mathrm{d}x \geq k_0^2 c_0 \int_\Omega |\hat{f}|^2 \,\mathrm{d}x,$$

which proves (3.45).

(ii) The first half of the proof is based on the idea of the proof of Theorem 4.8 (b) in [45]. We define $\hat{g} \in L^2(\Omega)$ by $\hat{g} = (\hat{f} + \sqrt{|q|}\, v_\alpha)\big|_\Omega$ and use the definition (3.33) of T to find

$$\langle T\hat{f}, \hat{f}\rangle_{L^2} = k_0^2 \int_\Omega \operatorname{sign}(q)\, |\hat{g}|^2 \, \mathrm{d}x - k_0^2 \int_\Omega \frac{q}{\sqrt{|q|}}\, \hat{g}\, \overline{v_\alpha} \, \mathrm{d}x. \tag{3.47}$$

Since v_α satisfies the equation

$$\Delta v_\alpha + k_0^2 v_\alpha = -\sqrt{|q|}\, T\hat{f} = -k_0^2 \frac{q}{\sqrt{|q|}}\, \hat{g} \tag{3.48}$$

in Π in the variational sense (cp. (3.34)), (3.47) can be written as

$$\langle T\hat{f}, \hat{f}\rangle_{L^2} = k_0^2 \int_\Omega \operatorname{sign}(q)\, |\hat{g}|^2 \, \mathrm{d}x + \int_\Omega (\Delta v_\alpha + k_0^2 v_\alpha)\, \overline{v_\alpha} \, \mathrm{d}x.$$

Now, we let $R = \{x \in \Pi : |x_3| < r\}$ with $r > 0$ such that $\overline{\Omega} \cap \Pi \subset R$. Applying the Green's identity (2.18) in Ω and afterwards in $R \backslash \overline{\Omega}$, we obtain

$$\begin{aligned}
&\langle T\hat{f}, \hat{f}\rangle_{L^2} \\
&= k_0^2 \int_\Omega \operatorname{sign}(q)\, |\hat{g}|^2 \, \mathrm{d}x + \int_\Omega (k_0^2 |v_\alpha|^2 - |\nabla v_\alpha|^2) \, \mathrm{d}x + \int_\Gamma \frac{\partial v_\alpha}{\partial \nu}\, \overline{v_\alpha} \, \mathrm{d}s \\
&= k_0^2 \int_\Omega \operatorname{sign}(q)\, |\hat{g}|^2 \, \mathrm{d}x + \int_R (k_0^2 |v_\alpha|^2 - |\nabla v_\alpha|^2) \, \mathrm{d}x + \int_{\partial R \cap \Pi} \frac{\partial v_\alpha}{\partial \nu}\, \overline{v_\alpha} \, \mathrm{d}s.
\end{aligned}$$

We point out that the boundary integrals have to be taken here only over $\Gamma = \partial\Omega \cap \Pi$ and $\partial R \cap \Pi$, respectively, since the contributions on $\overline{\Omega'} \cap \partial\Pi$ and $\overline{R} \cap \partial\Pi$ cancel out. Using the radiation condition (3.5), the orthogonality of $e^{\mathrm{i}\,\alpha_z \cdot x}$, $z \in Z$, as functions of $(x_1, x_2) \in (-\pi, \pi)^2$, and the fact that $\nu = \pm e_3$ on $\partial R \cap \Pi$, we arrive at

$$\begin{aligned}
\langle T\hat{f}, \hat{f}\rangle_{L^2} = k_0^2 \int_\Omega \operatorname{sign}(q)\, |\hat{g}|^2 \, \mathrm{d}x + \int_R (k_0^2 |v_\alpha|^2 - |\nabla v_\alpha|^2) \, \mathrm{d}x + \\
+ \mathrm{i} \int_{\partial R \cap \Pi} \sum_{z \in Z} \beta_z |v_z^\pm|^2 e^{-2\operatorname{Im}\beta_z x_3} \, \mathrm{d}s.
\end{aligned} \tag{3.49}$$

Finally, letting $r \to +\infty$ and noting that then in (3.49) the partial sum of the terms corresponding to $z \in Z$ with $\beta_z \in \mathrm{i}\,\mathbb{R}^+$ vanishes, we obtain

$$\begin{aligned}
\langle T\hat{f}, \hat{f}\rangle_{L^2} = k_0^2 \int_\Omega \operatorname{sign}(q)\, |\hat{g}|^2 \, \mathrm{d}x + \int_\Pi (k_0^2 |v_\alpha|^2 - |\nabla v_\alpha|^2) \, \mathrm{d}x + \\
+ \mathrm{i}\, 4\pi^2 \sum_{\beta_z \in \mathbb{R}^+} \beta_z |v_z^\pm|^2.
\end{aligned}$$

Under Assumptions 3.2, this yields

$$\operatorname{Im}\langle T\hat{f},\hat{f}\rangle_{L^2} = k_0^2 \int_\Omega \frac{\operatorname{Im} q}{|q|}|\hat{g}|^2\,\mathrm{d}x + 4\pi^2 \sum_{\beta_z\in\mathbb{R}^+} \beta_z |v_z^\pm|^2 \geq 0, \tag{3.50}$$

which is the assertion.

(iii) The short proof is in the same spirit as the proof of Theorem 5.12 (c) in [45]. Assume that there is no constant $c_2 > 0$ such that (3.46) holds for all $\hat{f} \in L^2(\Omega)$. Then we can find a sequence $\hat{f}_j$, $j \in \mathbb{N}$, with $\|\hat{f}_j\|_{L^2} = 1$ such that $\operatorname{Im}\langle T\hat{f},\hat{f}\rangle_{L^2} \to 0$. From (3.50) and the definition of $\hat{g}$ we conclude that $\hat{g}_j = (\hat{f}_j + \sqrt{|q|}\,v_{\alpha,j})|_\Omega \to 0$ in $L^2(\Omega)$, where $v_{\alpha,j}$ denotes the radiating solution to (3.31) with $\hat{f}$ replaced by $\hat{f}_j$. Thus, $v_{\alpha,j}$ is the radiating variational solution to

$$\Delta v_{\alpha,j} + k_0^2 v_{\alpha,j} = -k_0^2 \frac{q}{\sqrt{|q|}}\hat{f}_j - k_0^2 q v_{\alpha,j} = -k_0^2 \frac{q}{\sqrt{|q|}}\hat{g}_j$$

in Π, where the right-hand side is extended by zero into Ω^{ext}. From Proposition 3.3 we obtain that $v_{\alpha,j}$, $j \in \mathbb{N}$, converges to zero in $H^1_\alpha(\Omega)$ and therefore $\hat{f}_j \to 0$ in $L^2(\Omega)$. This contradicts $\|\hat{f}_j\|_{L^2} = 1$, the assertion is proven.

(iv) Let $\hat{f} \in \ker T$. Then (3.48) turns into the Helmholtz equation $\Delta v_\alpha + k_0^2 v_\alpha = 0$ in Π. This means that from the theoretical perspective the scattering medium is invisible for $\hat{f} \in \ker T$. We understand the Helmholtz equation in the variational sense, but recall that every variational solution to it is indeed a classical one. Now, we apply the representation formulas (3.20) and (3.21) with any bounded Lipschitz set $L \subset \Pi$. Since in the situation considered here the assumptions for both formulas are fulfilled, we combine them to conclude that v_α vanishes identically in Π. Then from

$$\begin{aligned} & T\hat{f} = k_0^2 \operatorname{sign}(q)\big(\hat{f} + \sqrt{|q|}\,v_\alpha\big)\big|_\Omega = 0 \\ \Longleftrightarrow\quad & \big(\hat{f} + \sqrt{|q|}\,v_\alpha\big)\big|_\Omega = 0 \end{aligned} \tag{3.51}$$

there follows $\hat{f} \equiv 0$, hence T is injective. □

We finally remark that the weighting factor $\sqrt{|q|}$ in the definition (3.29) of H_{Γ_s} is important for the coercivity of $\operatorname{Re}(e^{\mathrm{i}t}\widetilde{T})$ and $\operatorname{Im} T$. If e.g. the contrast q is continuous, then the coercivity could not be ensured without the weighting factor because of the resulting form of the operator T and the fact that q decays to zero in a neighborhood of Γ.

Chapter 4

The electromagnetic case

4.1 The direct problem

4.1.1 Problem formulation

Like in the acoustic case, we start regardless of regularity issues by introducing the basic model equations for the electromagnetic direct problem. These are the α-quasi-periodic time-harmonic Maxwell's equations

$$\operatorname{curl} \widetilde{H}_\alpha + \mathrm{i}\, \omega\, \varepsilon\, \widetilde{E}_\alpha = \sigma\, \widetilde{E}_\alpha, \qquad \operatorname{curl} \widetilde{E}_\alpha - \mathrm{i}\, \omega\, \mu\, \widetilde{H}_\alpha = 0,$$

where $\widetilde{H}_\alpha$ and $\widetilde{E}_\alpha$ represent an α-quasi-periodic magnetic and electric field, respectively. Again, we let the permeability μ be constant equal to its value μ_0 in vacuum, hence the relative permeability $\mu_r(x) = \mu(x)/\mu_0$ equals one in all of $\mathbb{R}^3$. We also recall that $\varepsilon_r(x) = \hat{\varepsilon}(x)/\varepsilon_0$ with $\hat{\varepsilon}(x) = \varepsilon(x) + \mathrm{i}\, \sigma(x)/\omega$ denotes the relative permittivity. Substituting in the above equations $\widetilde{H}_\alpha$ by $\mu_0^{-1/2} H_\alpha$ and $\widetilde{E}_\alpha$ by $\varepsilon_0^{-1/2} E_\alpha$, we obtain

$$\operatorname{curl} H_\alpha + \mathrm{i}\, k_0\, \varepsilon_r\, E_\alpha = 0, \qquad \operatorname{curl} E_\alpha - \mathrm{i}\, k_0\, H_\alpha = 0, \tag{4.1}$$

where $k_0 = \omega \sqrt{\varepsilon_0\, \mu_0}$ is the wave number in vacuum, being the background matter in the model. We assume at first that the equations (4.1) hold in the whole unit cell Π. If these equations hold in the classical sense then, due to the identity $\operatorname{div}\operatorname{curl} = 0$, they imply that $\operatorname{div}(\varepsilon_r E_\alpha) = 0$ and $\operatorname{div} H_\alpha = 0$. This means that for a constant permeability the magnetic field H_α is divergence-free (in the classical sense), whereas for the normal case of a non-constant complex permittivity $\hat{\varepsilon}$ the electric field E_α is not. Since this feature of H_α remains valid in a proper sense also for the notion of a weak (or variational) solution (H_α, E_α) to (4.1), which we discuss below, it is convenient to deal with the magnetic field rather than the electric field. Once H_α is found, E_α can be computed according to the first equation in (4.1). Utilizing the first in the second equation yields the second-order Maxwell's equation $\operatorname{curl}(\varepsilon_r^{-1}\operatorname{curl} H_\alpha) - k_0^2 H_\alpha = 0$ for the magnetic field in Π. The full direct problem for the electromagnetic case now reads in a first version: Given an incident α-quasi-periodic magnetic field H_α^i, compute the corresponding scattered magnetic field H_α^s as a solution to the equations

$$\operatorname{curl}\left(\frac{1}{\varepsilon_r}\operatorname{curl} H_\alpha\right) - k_0^2 H_\alpha = 0 \qquad \text{in } \Pi, \tag{4.2}$$

$$H_\alpha = H_\alpha^i + H_\alpha^s \qquad \text{in } \Pi, \tag{4.3}$$

$$[\nu \times H_\alpha]_\Gamma = 0, \tag{4.4}$$

$$[\nu \times E_\alpha]_\Gamma = 0 \Leftrightarrow [\nu \times (\varepsilon_r^{-1}\operatorname{curl} H_\alpha)]_\Gamma = 0. \tag{4.5}$$

Again, $[f]_\Gamma = f|_+ - f|_-$ denotes the jump of f across the boundary $\Gamma = \partial\Omega \cap \Pi$ and ν is the exterior unit normal vector to Ω. In analogy to the Rayleigh expansion (3.5) in acoustics, we require in addition that H_α^s obeys a representation of the form

$$H_\alpha^s(x) = \sum_{z\in Z} \operatorname{curl}\left(\widetilde{h}_z^\pm e^{\mathrm{i}(\alpha_z\cdot x \pm \beta_z x_3)}\right) = \sum_{z\in Z} h_z^\pm e^{\mathrm{i}(\alpha_z\cdot x \pm \beta_z x_3)} \qquad \text{in } R_\pm, \tag{4.6}$$

where $Z = \mathbb{Z}^2 \times \{0\}$, $\alpha_z = \alpha + z$, $\beta_z = \sqrt{k_0^2 - |\alpha_z|^2}$, and

$$h_z^\pm = \mathrm{i}\begin{pmatrix} \alpha_{z,1} \\ \alpha_{z,2} \\ \pm\beta_z \end{pmatrix} \times \widetilde{h}_z^\pm.$$

The series in (4.6) is assumed to converge uniformly on compact subsets of $R_+ \cup R_-$. Moreover, we require that $\beta_z \neq 0$ for all $z \in Z$. With $k_0 = \omega\sqrt{\varepsilon_0\mu_0}$, the associated *Rayleigh frequencies* form the set

$$\mathcal{E} = \{\omega \in \mathbb{R}^+ : \omega = (\varepsilon_0\mu_0)^{-1/2}|\alpha_z| \text{ for some } z \in Z\}.$$

The *Rayleigh expansion* (4.6) acts as the radiation condition for the Maxwell problem. Obviously, for given phase shift α and wave number k_0, the *Rayleigh coefficient vectors* $h_z^\pm \in \mathbb{C}^3$, $z \in Z$, completely characterize the scattered field H_α^s. A function which satisfies (4.6) is said to be *radiating* in the following. Together, the equations (4.2)–(4.6) model an α-quasi-periodic electromagnetic transmission problem, with transmission in Ω' (recall $\Omega = \Omega' \cap \Pi$). We now fix the class of incident fields and to this end let $\Gamma_i = \Gamma_{i,+} \cup \Gamma_{i,-}$ be a surface of the same type as in the acoustic case, cp. p. 30. We consider α-quasi-periodic incident fields

H^i_α which originate from a collection of point sources located on Γ_i. The point sources are magnetic dipoles here. Such a field solves the homogeneous second-order Maxwell's equation $\operatorname{curl}^2 H^i_\alpha - k_0^2 H^i_\alpha = 0$ in $\Pi\backslash\Gamma_i$ and is smooth there. For the scattered field H^s_α, we then obtain (formally) the equation

$$\operatorname{curl}\left(\frac{1}{\varepsilon_r}\operatorname{curl} H^s_\alpha\right) - k_0^2 H^s_\alpha = \operatorname{curl}(q \operatorname{curl} H^i_\alpha) \qquad \text{in } \Pi. \tag{4.7}$$

With (4.3) and the continuity of H^i_α across Γ, from the transmission conditions (4.4) and (4.5) we derive

$$[\nu \times H^s_\alpha]_\Gamma = 0 \qquad \text{and} \qquad [\nu \times (\varepsilon_r^{-1} \operatorname{curl} H^s_\alpha)]_\Gamma = -\nu \times (q \operatorname{curl} H^i_\alpha)\big|_-. \tag{4.8}$$

For the following problem treatment, we let the relative permittivity ε_r and the contrast $q = 1 - 1/\varepsilon_r$ be elements of $L^\infty(\Pi)$. However, we refer back to the explanation on p. 30 and make, in fact, the same implicit assumptions for ε_r and q as for the contrast in the acoustic case. We restrict also here to stating the transmission conditions at $\Gamma = \partial\Omega \cap \Pi$. Consequently, there is no classical solution to the equations (4.6)–(4.8) in general. We specify now a proper notion of a solution to this problem and comment on the formal rearranging leading to (4.7) and (4.8).

Similar to the approach in the acoustic case, we consider a slightly more abstract **direct problem**: Given a vectorial function f with support in $\overline{\Omega}$, find a radiating function v_α which satisfies

$$\operatorname{curl}\left(\frac{1}{\varepsilon_r}\operatorname{curl} v_\alpha\right) - k_0^2 v_\alpha = \operatorname{curl} f \qquad \text{in } \Pi, \tag{4.9}$$

together with the transmission conditions

$$[\gamma_t v_\alpha]_\Gamma = 0 \qquad \text{and} \qquad [\gamma_t(\varepsilon_r^{-1} \operatorname{curl} v_\alpha)]_\Gamma = -\gamma_{t,-} f. \tag{4.10}$$

Here, $[\gamma_t v_\alpha]_\Gamma$ denotes the jump of the tangential trace across Γ and is given by $[\gamma_t v_\alpha]_\Gamma = -\gamma_{t,+} v_\alpha|_\Gamma - \gamma_{t,-} v_\alpha|_\Gamma$, where $\gamma_{t,+}$ and $\gamma_{t,-}$ are the tangential trace operators for Ω^{ext} and Ω, respectively. The minus sign in front of $\gamma_{t,+}$ is due to the fact that $\gamma_{t,+}$ is the generalization of the operator $\nu \times \gamma_{D,+}(\cdot)$ on $\partial\Omega^{\text{ext}}$, where the normal vector ν points *into* rather than out of Ω on Γ (check against (4.8)).

Variational formulation Under the given assumptions, we understand (4.9) with (4.10) in the variational sense. For $f \in L^2(\Pi, \mathbb{C}^3)$, we seek a radiating function $v_\alpha \in H_{\alpha,\text{loc}}(\operatorname{curl}, \Pi)$ such that

$$\int_\Pi \left(\frac{1}{\varepsilon_r}\operatorname{curl} v_\alpha \cdot \operatorname{curl} \psi_{-\alpha} - k_0^2 v_\alpha \cdot \psi_{-\alpha}\right) \mathrm{d}x = \int_\Omega f \cdot \operatorname{curl} \psi_{-\alpha} \, \mathrm{d}x \tag{4.11}$$

is fulfilled for all $\psi_{-\alpha} \in H_{-\alpha}(\mathrm{curl},\Pi)$ with compact support with respect to the x_3-variable, meaning that there is a compact set $M \subset \overline{\Pi}$ such that $\psi_{-\alpha}((x_1,x_2,\cdot))$ is (essentially) supported in M for all $(x_1,x_2) \in (-\pi,\pi)^2$. We recall that the α-quasi-periodic extension of a function in $H_{\alpha,\mathrm{loc}}(\mathrm{curl},\Pi)$ is $H_\alpha(\mathrm{curl})$-regular across the boundary $\partial\Pi$, cf. the definition of the space $H_{\alpha,\mathrm{loc}}(\mathrm{curl},S)$ on p. 27. With the spaces chosen above, (4.11) is well-defined and v_α naturally meets the first transmission condition in (4.10). However, in order to explain the relation between (4.11) and (4.9) with (4.10), we have to clarify some points. First, we note that for $f \in L^2(\Pi,\mathbb{C}^3)$ the right-hand side of the second transmission condition in (4.10) is certainly not well-defined. Even more, the left-hand side of the condition as well as the equation (4.9) in the actual variational sense are not well-defined in general. To see this, one should be aware that for the derivation of the variational formulation the Green's identity (2.25) has to be applied, but neither $\varepsilon_r^{-1}\,\mathrm{curl}\,v_\alpha$ nor f is $H(\mathrm{curl})$-regular. The equation (4.11) is obtained by a formal application of this identity to (4.9). Nevertheless, it can be justified rigorously as follows. We go back to our starting point (4.1) and note that it is reasonable to require that the physical total fields H_α and $E_\alpha = -(\mathrm{i}k_0\,\varepsilon_r)^{-1}\,\mathrm{curl}\,H_\alpha$ are $H_\alpha(\mathrm{curl})$-regular, in order for the Maxwell's equations (4.1) to be fulfilled in a proper sense. Also, the physical incident field H_α^i as well as $\mathrm{curl}\,H_\alpha^i$ are $H_\alpha(\mathrm{curl})$-regular since H_α^i is smooth in its domain of definition. Hence, under consideration of the regularity issues, the equations in (4.7) and (4.8) are stated in a proper form as

$$\mathrm{curl}\left(\frac{1}{\varepsilon_r}\mathrm{curl}\,H_\alpha - \mathrm{curl}\,H_\alpha^i\right) - k_0^2 H_\alpha^s = 0 \qquad \text{in } \Pi \tag{4.7'}$$

in the variational sense, together with the transmission conditions

$$\left.\begin{aligned} [\gamma_t H_\alpha^s]_\Gamma &= 0, \\ [\gamma_t(\varepsilon_r^{-1}\,\mathrm{curl}\,H_\alpha - \mathrm{curl}\,H_\alpha^i)]_\Gamma &= 0 \\ \Longleftrightarrow \quad \gamma_{t,+}(\mathrm{curl}\,H_\alpha^s) + \gamma_{t,-}(\varepsilon_r^{-1}\,\mathrm{curl}\,H_\alpha) &= \gamma_{t,-}(\mathrm{curl}\,H_\alpha^i). \end{aligned}\right\} \tag{4.8'}$$

Applying the Green's identity (2.25) to (4.7′) separately in the domains Ω and Ω^{ext}, the variational form of (4.7′) is found to be

$$\int_\Pi \left(\left(\frac{1}{\varepsilon_r}\mathrm{curl}\,H_\alpha - \mathrm{curl}\,H_\alpha^i\right)\cdot \mathrm{curl}\,\psi_{-\alpha} - k_0^2 H_\alpha^s \cdot \psi_{-\alpha}\right)\mathrm{d}x - \\ - \int_\Gamma \left[\gamma_t(\varepsilon_r^{-1}\,\mathrm{curl}\,H_\alpha - \mathrm{curl}\,H_\alpha^i)\cdot \gamma_T\psi_{-\alpha}\right]_\Gamma \mathrm{d}s = 0. \tag{4.12}$$

Here, we have used that the integral contributions on $\partial\Pi$ cancel out. By the second transmission condition in (4.8′) and the fact that $\psi_{-\alpha} \in H_{-\alpha}(\mathrm{curl},\Pi)$ implies

$[\gamma_t \psi_{-\alpha}]_\Gamma = 0$ and likewise $[\gamma_T \psi_{-\alpha}]_\Gamma = 0$, the boundary integral term in (4.12) vanishes. On the other hand, if this boundary integral is required to vanish for all $\psi_{-\alpha} \in H_{-\alpha}(\text{curl}, \Pi)$ then this implies $[\gamma_t(\varepsilon_r^{-1} \,\text{curl}\, H_\alpha - \text{curl}\, H_\alpha^i)]_\Gamma = 0$. The formulation (4.11) is based on the observation that the remaining volume integral on the left-hand side of (4.12) stays well-defined for $H_\alpha \in H_{\alpha,\text{loc}}(\text{curl}, \Pi)$, without requiring that $\varepsilon_r^{-1} \,\text{curl}\, H_\alpha$ lies in $H_{\alpha,\text{loc}}(\text{curl}, \Pi)$. From this more general perspective, H_α is not understood as a magnetic field anymore. Now, one can rearrange terms and, replacing $H_\alpha^s = H_\alpha - H_\alpha^i$ by v_α and $(1 - \varepsilon_r^{-1}) \,\text{curl}\, H_\alpha^i = q \,\text{curl}\, H_\alpha^i$ by f (as done before at an early stage), one arrives at (4.11). This derivation shows that only for certain sources $f \in L^2(\Pi, \mathbb{C}^3)$ and for certain radiating solutions $v_\alpha \in H_{\alpha,\text{loc}}(\text{curl}, \Pi)$ to (4.11), these v_α can be reinterpreted as variational solutions to the physical problem (4.7′) with (4.8′), where v_α stands for H_α^s, f/q stands for $\text{curl}\, H_\alpha^i$ in Ω, and $\varepsilon_r^{-1} \,\text{curl}\, H_\alpha - \text{curl}\, H_\alpha^i$ equals $\text{curl}\, v_\alpha$ in Ω^{ext}. In this case, (4.11) incorporates the second transmission condition in (4.8′) as explained above. Keeping this relation between (4.11) and the physical problem in mind, we continue to use the convenient 'schematic' formulation by (4.9) and (4.10).

Existence and uniqueness of a variational solution to the direct problem can be shown by a similar procedure as outlined for the acoustic case at the end of Subsection 3.1.1. These and more results for the electromagnetic direct problem, for the medium type considered here and a plane wave incidence, are obtained in the articles [8, 63]. We finally comment on the regularity of a radiating variational solution v_α to (4.9) with (4.10). For constant permeability μ, as we assume it to be the case with $\mu \equiv \mu_0$, the α-quasi-periodic extension of v_α, which we identify with v_α, is divergence-free in all of $\mathbb{R}^3$ in the weak sense. To show this, one uses the Green's identity (2.26) and the fact that for constant μ the normal trace $\gamma_n v_\alpha$ of v_α does not jump in $\mathbb{R}^3$ (cp. Subsection 1.2.2 in [55]). Hence, v_α lies in the space

$$H_\alpha^\diamond(\Pi) = \{u_\alpha \in H_{\alpha,\text{loc}}(\text{curl}, \Pi) : \text{div}\, u_\alpha \in L^2_{\text{loc}}(\mathbb{R}^3)\}$$

and so in fact is H_α^1-regular on every compact subset of $\mathbb{R}^3$, cf. Corollary 2.10 in [28]. In Ω^{ext}, i.e. outside the supports of the contrast q and the source f in Π, the equation (4.9) reduces to the homogeneous equation $\text{curl}^2 v_\alpha - k_0^2 v_\alpha = 0$. The classical differential operators $\mathcal{D}_j$, $j \in \{1, 2, 3\}$, which are uniquely determined by

$$\begin{pmatrix} \mathcal{D}_1 v_{\alpha,1} \\ \mathcal{D}_2 v_{\alpha,2} \\ \mathcal{D}_3 v_{\alpha,3} \end{pmatrix} = \text{curl}^2 v_\alpha - k_0^2 v_\alpha$$

are not strictly elliptic. However, using the identities $\operatorname{div}\operatorname{curl} = 0$ and $\operatorname{curl}^2 = -\Delta + \operatorname{grad}\operatorname{div}$, one finds that the homogeneous second-order Maxwell's equation is equivalent to the system

$$\left.\begin{aligned} \Delta v_\alpha + k_0^2 v_\alpha &= 0 \\ \operatorname{div} v_\alpha &= 0 \end{aligned}\right\}, \tag{4.13}$$

where the Laplace operator Δ is meant componentwise. Since this equivalence holds in the classical sense, so it does in the weak sense. Now, with $\Delta + k_0^2\,\mathrm{id}$ being strictly elliptic componentwise, we can apply Corollary 8.11 in [27] to conclude that every weak solution to $\operatorname{curl}^2 v_\alpha - k_0^2 v_\alpha = 0$ is a classical one. According to Theorem 3.5 in [16], v_α is even analytic in Ω^{ext}. Thus, in particular, the Rayleigh expansion (4.6) is well-defined for v_α.

4.1.2 The Green's tensor and representation theorems

Like for the treatment of the acoustic problem, we choose an integral equation approach. In the electromagnetic case, this necessitates the α-quasi-periodic Green's tensor for the Maxwell operator $\operatorname{curl}^2 - k_0^2\,\mathrm{id}$. It is simple to verify that this tensor is given by

$$\mathfrak{G}_\alpha(y,x) = G_\alpha(y,x)\,\mathbb{I}_{3\times3} + k_0^{-2}\operatorname{grad}_x\operatorname{div}_x(G_\alpha(y,x)\,\mathbb{I}_{3\times3}) \tag{4.14}$$

where $x, y \in \Pi$ with $x \neq y$ (see [50]). Here, G_α again denotes the α-quasi-periodic scalar Green's function for the Helmholtz operator, $\mathbb{I}_{3\times3}$ is the identity in $\mathbb{C}^{3\times3}$, and div is meant columnwise, grad componentwise. As a consequence of the relation (3.14) for the scalar Green's function and the identities $\operatorname{curl}^2 = -\Delta + \operatorname{grad}\operatorname{div}$ and $\operatorname{curl}\operatorname{grad} = 0$, $\mathfrak{G}_\alpha$ as a distribution formally satisfies

$$\operatorname{curl}_x^2 \mathfrak{G}_\alpha(y,x) - k_0^2\,\mathfrak{G}_\alpha(y,x) = \sum_{z\in Z} e^{2\pi\mathrm{i}\,\alpha\cdot z}\delta_{y+2\pi z}(x)\,\mathbb{I}_{3\times3}, \tag{4.15}$$

which implies

$$\begin{aligned} \operatorname{curl}_x^2 \mathfrak{G}_\alpha(y,x) - k_0^2\,\mathfrak{G}_\alpha(y,x) &= \delta_y(x)\,\mathbb{I}_{3\times3} \qquad \text{and} \\ \operatorname{curl}_y^2 \mathfrak{G}_\alpha(y,x) - k_0^2\,\mathfrak{G}_\alpha(y,x) &= \delta_x(y)\,\mathbb{I}_{3\times3} \end{aligned}$$

for $x, y \in \Pi$. The tensor $\mathfrak{G}_\alpha$ maps (y,x) to a symmetric matrix and there holds $\mathfrak{G}_\alpha(y,x) = \mathfrak{G}_{-\alpha}(x,y)$. Moreover, it is easy to see that $\mathfrak{G}_\alpha$ fulfills radiation conditions of the forms

$$\mathfrak{G}_\alpha(y,x) = \operatorname{curl}_x \sum_{z\in Z} \widetilde{G}_z^{\pm}(y)\, e^{\mathrm{i}(\alpha_z\cdot x \pm \beta_z x_3)}, \qquad x_3 \gtrless y_3, \qquad \text{and} \tag{4.16}$$

$$\mathfrak{G}_\alpha(y,x) = \operatorname{curl}_y \sum_{z\in Z} \widetilde{G}_z^{\pm}(x)\, e^{-\mathrm{i}(\alpha_z\cdot y \pm \beta_z y_3)}, \qquad x_3 \gtrless y_3, \tag{4.17}$$

for fixed y and fixed x, respectively. Here, $\widetilde{G}_z^{\pm}(y)$ and $\widetilde{G}_z^{\pm}(x)$, $z \in Z$, are coefficient matrices and curl is meant columnwise. For a clearer view of (4.16), one should consider that the columns of $\mathfrak{G}_\alpha(y,\cdot)$ satisfy the Rayleigh radiation condition (4.6). A similar statement applies to (4.17) for $\mathfrak{G}_\alpha(\cdot,x)$.

Representation theorems

Now, having introduced the Green's tensor, we can establish representations for α-quasi-periodic vector fields in the interior and the exterior of some Lipschitz set. These representations can be proven to be equivalent to the well-known Stratton-Chu formulas stated for our setting, see Section 9.2 in [55] and, for smooth domains, Section 6.2 in [17]. The following theorem is the counterpart of Theorem 3.1.

Theorem 4.1. *Let $L \subset \Pi$ be a bounded Lipschitz set such that L_{per} is also Lipschitz and $L_c = \overline{L_{per}} \cap \overline{\Pi}$ is not degenerate, according to Definition 2.1.*

(i) Assume $v_\alpha \in H_\alpha(\operatorname{curl}^2, L)$. Then there holds

$$\int_{\partial L} \Big(\mathfrak{G}_\alpha(y,x)\big(\nu(y)\times \operatorname{curl} v_\alpha(y)\big) - \operatorname{curl}_y \mathfrak{G}_\alpha(y,x)\big(\nu(y)\times v_\alpha(y)\big)\Big)\mathrm{d}s(y) -$$
$$- \int_L \mathfrak{G}_\alpha(y,x)\big(\operatorname{curl}^2 v_\alpha(y) - k_0^2 v_\alpha(y)\big)\mathrm{d}y = \begin{cases} -v_\alpha(x), & x \in L \\ 0, & x \in \Pi\backslash\overline{L} \end{cases} \tag{4.18}$$

for almost all $x \in \Pi$. If $v_\alpha \in C_\alpha^2(\overline{L}, \mathbb{C}^3)$, then (4.18) holds for all $x \in \Pi$.

(ii) Assume $v_\alpha \in H_{\alpha,loc}(\operatorname{curl}, \Pi\backslash\overline{L})$ is a radiating solution to the homogeneous Maxwell's equation $\operatorname{curl}^2 v_\alpha - k_0^2 v_\alpha = 0$ in $\Pi\backslash\overline{L}$. Then v_α is represented by

$$\int_{\partial L} \Big(\mathfrak{G}_\alpha(y,x)\big(\nu(y)\times \operatorname{curl} v_\alpha(y)\big) - \operatorname{curl}_y \mathfrak{G}_\alpha(y,x)\big(\nu(y)\times v_\alpha(y)\big)\Big)\mathrm{d}s(y)$$
$$= \begin{cases} 0, & x \in L \\ v_\alpha(x), & x \in \Pi\backslash\overline{L} \end{cases}. \tag{4.19}$$

The first integrals on the left-hand sides of (4.18) and (4.19) can be rewritten componentwise as

$$\int_{\partial L}\Big(\mathfrak{G}_\alpha(y,x)^{(j)}\cdot\big(\nu(y)\times\operatorname{curl}v_\alpha(y)\big)-v_\alpha(y)\cdot\big(\nu(y)\times\operatorname{curl}_y\mathfrak{G}_\alpha(y,x)^{(j)}\big)\Big)\,\mathrm{ds}(y) \tag{4.20}$$

where $\mathfrak{G}_\alpha^{(j)}$ *denotes the j-th column of* $\mathfrak{G}_\alpha$.

Proof. We only prove the first case in (4.18). To start, let $x\in\Pi$ and $L_\rho=\{y\in L:|x-y|\geq\rho\}$ for some $\rho>0$. For the ease of notation, in all integrals below, hidden dependencies and derivatives are meant with respect to the variable y. By using twice the identity (2.24), for $u\in H(\operatorname{curl}^2,L_\rho)$ and v with $\operatorname{curl}v\in H^1(L_\rho,\mathbb{C}^3)$ we obtain

$$\int_{\partial L_\rho}\big((\nu\times\operatorname{curl}u)\cdot v+(\nu\times u)\cdot\operatorname{curl}v\big)\mathrm{ds}(y)=\int_{L_\rho}(\operatorname{curl}^2u\cdot v-\operatorname{curl}^2v\cdot u)\,\mathrm{d}y.$$

Now, we insert v_α for u and $\mathfrak{G}_\alpha(\cdot,x)^{(j)}$ for v and get

$$\begin{aligned}&\int_{\partial L_\rho}\big(\mathfrak{G}_\alpha(\cdot,x)^{(j)}\cdot(\nu\times\operatorname{curl}v_\alpha)+\operatorname{curl}_y\mathfrak{G}_\alpha(\cdot,x)^{(j)}\cdot(\nu\times v_\alpha)\big)\mathrm{ds}(y)\\&=\int_{L_\rho}\big(\operatorname{curl}^2v_\alpha\cdot\mathfrak{G}_\alpha(\cdot,x)^{(j)}-\operatorname{curl}_y^2\mathfrak{G}_\alpha(\cdot,x)^{(j)}\cdot v_\alpha\big)\mathrm{d}y\\&=\int_{L_\rho}\mathfrak{G}_\alpha(\cdot,x)^{(j)}\cdot(\operatorname{curl}^2v_\alpha-k_0^2v_\alpha)\,\mathrm{d}y\end{aligned}$$

with $x\in\Pi$. Moreover, using matrix notation and exploiting that $\mathfrak{G}_\alpha(y,x)$ is symmetric and $\operatorname{curl}_y\mathfrak{G}_\alpha(y,x)$ is skew symmetric, we obtain

$$\int_{\partial L_\rho}\big(\mathfrak{G}_\alpha(\cdot,x)\,(\nu\times\operatorname{curl}v_\alpha)-\operatorname{curl}_y\mathfrak{G}_\alpha(\cdot,x)\,(\nu\times v_\alpha)\big)\mathrm{ds}(y)$$
$$=\int_{L_\rho}\mathfrak{G}_\alpha(\cdot,x)\,(\operatorname{curl}^2v_\alpha-k_0^2v_\alpha)\,\mathrm{d}y \tag{4.21}$$

with $x\in\Pi$. With the identities

$$\begin{aligned}\operatorname{curl}_x\mathfrak{G}_\alpha(y,x)&=-\operatorname{curl}_y\mathfrak{G}_\alpha(y,x),\\\operatorname{curl}_x(\mathfrak{G}_\alpha(y,x)\,h(y))&=\operatorname{curl}_x(G_\alpha(y,x)\,h(y)),\qquad\text{and}\\\operatorname{curl}_x\operatorname{curl}_x(G_\alpha(y,x)\,h(y))&=k_0^2\,\mathfrak{G}_\alpha(y,x)\,h(y)\end{aligned}$$

for $x \neq y$ it is a straightforward computation to show that the left-hand side of (4.21) equals

$$\operatorname{curl} \int_{\partial L_\rho} (\nu \times v_\alpha)\, G_\alpha(\cdot,x)\, \mathrm{ds}(y) + \frac{1}{k_0^2} \operatorname{curl}^2 \int_{\partial L_\rho} (\nu \times \operatorname{curl} v_\alpha)\, G_\alpha(\cdot,x)\, \mathrm{ds}(y). \tag{4.22}$$

Let now $v_\alpha \in C^2_\alpha(\overline{L}, \mathbb{C}^3)$. Then, by the arguments on p. 158f. in [17] and those behind Theorem 3.1, for $x \in L$ the expression (4.22) turns into

$$v_\alpha(x) + \operatorname{curl} \int_{\partial L} (\nu \times v_\alpha)\, G_\alpha(\cdot,x)\, \mathrm{ds}(y) + \frac{1}{k_0^2} \operatorname{curl}^2 \int_{\partial L} (\nu \times \operatorname{curl} v_\alpha)\, G_\alpha(\cdot,x)\, \mathrm{ds}(y)$$

in the limit of $\rho \to 0$. Hence, if $v_\alpha \in C^2_\alpha(\overline{L}, \mathbb{C}^3)$ is a solution to $\operatorname{curl}^2 v_\alpha - k_0^2 v_\alpha = 0$ in L, from (4.21) we recover the Stratton-Chu formula

$$\begin{aligned} -v_\alpha(x) = \operatorname{curl} \int_{\partial L} (\nu \times v_\alpha)\, G_\alpha(\cdot,x)\, \mathrm{ds}(y) - \\ + \frac{1}{k_0^2} \operatorname{curl}^2 \int_{\partial L} (\nu \times \operatorname{curl} v_\alpha)\, G_\alpha(\cdot,x)\, \mathrm{ds}(y) \end{aligned}$$

for $x \in L$. Under sufficient regularity of the field v_α (as given) it is valid to consider Lipschitz domains here instead of C^2-domains as in Theorem 6.2 in [17], cf. Theorems 3.19 and 9.2 in [55]. Based on (4.21), we arrive at

$$\begin{aligned} -v_\alpha(x) = \int_{\partial L} \big(\mathfrak{G}_\alpha(\cdot,x)\,(\nu \times \operatorname{curl} v_\alpha) - \operatorname{curl}_y \mathfrak{G}_\alpha(\cdot,x)\,(\nu \times v_\alpha)\big)\, \mathrm{ds}(y) - \\ - \int_L \mathfrak{G}_\alpha(\cdot,x)\,(\operatorname{curl}^2 v_\alpha - k_0^2 v_\alpha)\, \mathrm{d}y \end{aligned} \tag{4.23}$$

for $v_\alpha \in C^2_\alpha(\overline{L}, \mathbb{C}^3)$ and $x \in L$. The denseness of $C^2_\alpha(\overline{L}, \mathbb{C}^3)$ in $H_\alpha(\operatorname{curl}^2, L)$ with respect to the norm of the latter finally leads to the asserted weak form in (4.18). The componentwise reformulation (4.20) of the first integral in (4.23) is easily seen by the symmetry of $\mathfrak{G}_\alpha(y,x)$, the skew symmetry of $\operatorname{curl}_y \mathfrak{G}_\alpha(y,x)$, and the vector identity $a \cdot (b \times c) = -c \cdot (b \times a)$ for $a, b, c \in \mathbb{C}^3$. In the proofs of the second case of (4.18) and the identities in (4.19), the radiation conditions (4.6) for v_α and (4.17) for $\mathfrak{G}_\alpha(\cdot,x)$ are used, cp. the proofs of Theorems 4.1 and 4.5 in [16]. □

4.1.3 The near field operator

We are now prepared to introduce the *near field operator* for the electromagnetic case, which will be our main object of interest in the following. Again, the computation of this operator is related to the solution of the direct problem in Ω^{ext}.

The incidence surface Γ_i and the measurement surface Γ_s are assumed to have the same form as in the acoustic case, i.e.

$$\Gamma_i = \Gamma_{i,+} \cup \Gamma_{i,-} \qquad \text{and} \qquad \Gamma_s = \Gamma_{s,+} \cup \Gamma_{s,-},$$

where $\Gamma_{i,\pm} \subset \overline{R_\pm} \cap \Pi$ and $\Gamma_{s,\pm} \subseteq \Gamma_\pm$ are flat surfaces with non-empty relative interiors in the planes which contain them, see Figure 3.1 on p. 36. We remind the reader that we consider α-quasi-periodic incident fields $\widetilde{u}^i_\alpha$ which originate from magnetic dipoles on Γ_i. Precisely, we let

$$\widetilde{u}^i_\alpha(x) = \int_{\Gamma_i} \mathfrak{G}_\alpha(y,x)\,\phi(y)\,\mathrm{ds}(y), \qquad x \in \Pi \backslash \Gamma_i, \tag{4.24}$$

where $\phi(y)$ is the vectorial moment of a dipole at $y \in \Gamma_i$, represented by $\mathfrak{G}_\alpha(y,\cdot)$. Due to the superposition principle, this field generates the scattered field

$$\widetilde{u}^s_\alpha(x) = \int_{\Gamma_i} \widetilde{u}^s_{p,\alpha}(x,y)\,\phi(y)\,\mathrm{ds}(y), \qquad x \in \Pi, \tag{4.25}$$

where $\widetilde{u}^s_{p,\alpha}(\cdot,y)$ is the scattering response to the field of the single dipole at $y \in \Gamma_i$. The associated *near field operator* $\widetilde{M} : L^2(\Gamma_i,\mathbb{C}^3) \to L^2(\Gamma_s,\mathbb{C}^3)$ reads

$$(\widetilde{\mathcal{M}}\phi)(x) = \int_{\Gamma_i} \widetilde{u}^s_{p,\alpha}(x,y)\,\phi(y)\,\mathrm{ds}(y), \qquad x \in \Gamma_s. \tag{4.26}$$

However, since the surfaces Γ_i and Γ_s are not required to coincide, the function spaces do not suit an adaptation of the Factorization Method. Using the same technique as in the acoustic case, we resolve this problem by means of the auxiliary near field operator $\mathcal{M} : L^2(\Gamma_s,\mathbb{C}^3) \to L^2(\Gamma_s,\mathbb{C}^3)$ defined by

$$(\mathcal{M}\varphi)(x) = \int_{\Gamma_s} u^s_{p,\alpha}(x,y)\,\varphi(y)\,\mathrm{ds}(y), \qquad x \in \Gamma_s, \tag{4.27}$$

where $u^s_{p,\alpha}(\cdot,y)$ stands for the response to the field of a complex conjugate dipole at $y \in \Gamma_s$, modeled by $\overline{\mathfrak{G}_{-\alpha}(y,\cdot)}$. It is essential here that still we can construct an approximation for $\mathcal{M}$ from the given moment function ϕ and measurements of the scattered field $\widetilde{u}^s_\alpha$ on Γ_s, caused by the physical incidence $\widetilde{u}^i_\alpha$. We explain this in detail in Chapter 5.

4.2 The inverse problem

We are interested in the following **inverse problem**: Given the scattered fields $\widetilde{u}^s_\alpha$ on Γ_s for all moment functions $\phi \in L^2(\Gamma_i,\mathbb{C}^3)$ (and a single fixed wave number k_0), determine the support of the contrast q!

Again, we develop a variant of the Factorization Method in order to solve this problem. The procedure resembles that for the acoustic case.

4.2.1 Factorization of the near field operator

Since we described the general idea of the Factorization Method at the beginning of Section 3.2, we directly proceed with a suitable factorization of the auxiliary near field operator $\mathcal{M}$ defined in (4.27). In the remainder of the section, we make similar general assumptions as in the acoustic case. We collect them in

Assumptions 4.2.

- The sets Ω' and $\Omega = \Omega' \cap \Pi$ are Lipschitz, and the set $\Omega_c = \overline{\Omega_{\text{per}}} \cap \overline{\Pi}$ is not degenerate.
- The connected components of Ω' are simply connected.
- The relative permittivity ε_r lies in $L^\infty(\Pi)$ and satisfies
  (i) $\varepsilon_r = 1 \Leftrightarrow q = 1 - 1/\varepsilon_r = 0$ almost everywhere (a.e.) in Ω^{ext},
  (ii) $\operatorname{Re}\varepsilon_r \geq c_0$ a.e. in Ω for some constant $c_0 > 0$,
  (iii) $\operatorname{Im}\varepsilon_r \geq 0 \Leftrightarrow \operatorname{Im} q \geq 0$ a.e. in Ω,
  (iv) $|\varepsilon_r - 1|$ is locally bounded from below in Ω, i.e. for every compact subset $S \subset \Omega$ there is a constant $c_S > 0$ such that $|\varepsilon_r - 1| \geq c_S$ a.e. in S.
- The direct problem, defined on p. 53, is uniquely solvable.

We note that the assumptions on ε_r imply in particular that $q \in L^\infty(\Pi)$. In the style of the artificial incident field

$$u^i_\alpha(x) = \int_{\Gamma_s} \overline{\mathfrak{G}_{-\alpha}(y,x)}\, \varphi(y)\, \mathrm{d}s(y), \qquad x \in \Pi \backslash \Gamma_s,$$

which underlies the near field operator $\mathcal{M}$, we define the integral operator $\mathcal{H}_{\Gamma_s} : L^2(\Gamma_s, \mathbb{C}^3) \to L^2(\Omega, \mathbb{C}^3)$ by

$$(\mathcal{H}_{\Gamma_s}\varphi)(x) = \sqrt{|q(x)|}\, \operatorname{curl} \int_{\Gamma_s} \overline{\mathfrak{G}_{-\alpha}(y,x)}\, \varphi(y)\, \mathrm{d}s(y), \qquad x \in \Omega. \tag{4.28}$$

The adjoint $\mathcal{H}^*_{\Gamma_s} : L^2(\Omega, \mathbb{C}^3) \to L^2(\Gamma_s, \mathbb{C}^3)$ of $\mathcal{H}_{\Gamma_s}$ is straightforwardly shown to read

$$(\mathcal{H}^*_{\Gamma_s} g)(x) = \operatorname{curl} \int_\Omega \mathfrak{G}_{-\alpha}(x,y)\, g(y)\, \sqrt{|q(y)|}\, \mathrm{d}y, \qquad x \in \Gamma_s. \tag{4.29}$$

Analog to the acoustic case, we define the *solution operator* $\mathcal{G} : L^2(\Omega, \mathbb{C}^3) \to L^2(\Gamma_s, \mathbb{C}^3)$ which maps $\hat{f} \in L^2(\Omega, \mathbb{C}^3)$ to $v_\alpha|_{\Gamma_s}$ where v_α radiates according to (4.6) and solves

$$\int_\Pi \left(\frac{1}{\varepsilon_r} \operatorname{curl} v_\alpha \cdot \operatorname{curl} \psi_{-\alpha} - k_0^2 v_\alpha \cdot \psi_{-\alpha} \right) \mathrm{d}x = \int_\Omega \frac{q}{\sqrt{|q|}} \hat{f} \cdot \operatorname{curl} \psi_{-\alpha} \, \mathrm{d}x$$

or, equivalently,

$$\int_\Pi (\operatorname{curl} v_\alpha \cdot \operatorname{curl} \psi_{-\alpha} - k_0^2 v_\alpha \cdot \psi_{-\alpha}) \, \mathrm{d}x = \int_\Omega \frac{q}{\sqrt{|q|}} (\hat{f} + \sqrt{|q|} \operatorname{curl} v_\alpha) \cdot \operatorname{curl} \psi_{-\alpha} \, \mathrm{d}x \quad (4.30)$$

for all $\psi_{-\alpha} \in H_{-\alpha}(\operatorname{curl}, \Pi)$ with compact support in the x_3-dimension. The equation (4.30) is equivalent to the variational formulation (4.11) of the direct problem except that here $\hat{f}$ plays the role of $\sqrt{|q|}/q f|_\Omega$ in (4.11). Recalling the initial substitution of the source term $q \operatorname{curl} H^i_\alpha$ from (4.7) by f in (4.9), the requirement $\hat{f} \in L^2(\Omega, \mathbb{C}^3)$ appears reasonable. By Assumptions 4.2, there is a unique radiating solution to (4.30), thus the operator $\mathcal{G}$ is well-defined. From the above definitions, it follows that the near field operator $\mathcal{M}$ can be factorized by

$$\mathcal{M} = \mathcal{G} \mathcal{H}_{\Gamma_s}. \quad (4.31)$$

Moreover, motivated by the right-hand side of (4.30), we define the operator $\mathcal{T} : L^2(\Omega, \mathbb{C}^3) \to L^2(\Omega, \mathbb{C}^3)$ by

$$\mathcal{T} \hat{f} = \operatorname{sign}(q) \big(\hat{f} + \sqrt{|q|} \operatorname{curl} v_\alpha \big)\big|_\Omega \quad (4.32)$$

where $\operatorname{sign}(z) = z/|z|$ for $z \in \mathbb{C}$ and $v_\alpha \in H_{\alpha,\mathrm{loc}}(\operatorname{curl}, \Pi)$ satisfies (4.30) with the source $\hat{f}$ given as the argument of $\mathcal{T}$. Then, (4.30) is recognized as the variational form of

$$\operatorname{curl}^2 v_\alpha - k_0^2 v_\alpha = \operatorname{curl} \left(\frac{q}{\sqrt{|q|}} \hat{f} + q \operatorname{curl} v_\alpha \right) = \operatorname{curl} \big(\sqrt{|q|} \, \mathcal{T} \hat{f} \big) \quad (4.33)$$

in Π, where the right-hand side is extended by zero into Ω^{ext}. As a key step in our approach, we now prove that (4.30) can be equivalently formulated as an integro-differential equation. This will allow us to refine the factorization (4.31) of $\mathcal{M}$ and to deduce the so-called *electromagnetic Lippmann-Schwinger equation*. The central ingredient is the following result, cp. Lemma 5.2 in [45].

Proposition 4.3. *Let $\mathcal{W}$ be the volume potential operator defined by*

$$(\mathcal{W}g)(x) = \operatorname{curl} \int_\Omega \mathfrak{G}_{-\alpha}(x,y)\, g(y)\, \mathrm{d}y \tag{4.34}$$

with $g \in L^2(\Omega,\mathbb{C}^3)$ and $x \in \Pi$. We consider the potential $w_\alpha = \mathcal{W}g$ for some density g.

(i) Assume for now that the connected components of $\Gamma = \partial\Omega \cap \Pi = \partial\Omega' \cap \Pi$ are C^2-regular. (Clearly, this holds in particular if $\partial\Omega' \in C^2$.) For Hölder continuous densities $g \in C^{1,\gamma}(\overline{\Omega},\mathbb{C}^3)$ with $0 < \gamma \leq 1$ the potential $w_\alpha = \mathcal{W}g$ is in $C^2_\alpha(\Pi\backslash\Gamma,\mathbb{C}^3) \cap C_\alpha(\Pi,\mathbb{C}^3)$ and a classical radiating solution to the transmission problem

$$\operatorname{curl}^2 w_\alpha - k_0^2 w_\alpha = \operatorname{curl} g \qquad \text{in } \Pi\backslash\Gamma, \tag{4.35}$$

$$[\gamma_t w_\alpha]_\Gamma = 0, \tag{4.36}$$

$$[\gamma_t(\operatorname{curl} w_\alpha)]_\Gamma = g \times \nu. \tag{4.37}$$

In (4.35), we have extended the right-hand side by zero into Ω^{ext}.

(ii) Let again Assumptions 4.2 hold and Ω be just Lipschitz. For densities $g \in L^2(\Omega,\mathbb{C}^3)$, the potential w_α is a radiating variational solution to the equation $\operatorname{curl}^2 w_\alpha - k_0^2 w_\alpha = \operatorname{curl} g$ in Π, where the right-hand side is extended by zero into Ω^{ext}. This means that w_α lies in $H_{\alpha,loc}(\operatorname{curl},\Pi)$ and satisfies

$$\int_\Pi (\operatorname{curl} w_\alpha \cdot \operatorname{curl} \psi_{-\alpha} - k_0^2 w_\alpha \cdot \psi_{-\alpha})\, \mathrm{d}x = \int_\Omega g \cdot \operatorname{curl} \psi_{-\alpha}\, \mathrm{d}x \tag{4.38}$$

for all $\psi_{-\alpha} \in H_{-\alpha}(\operatorname{curl},\Pi)$ with compact support in the x_3-dimension.

(iii) The mapping of g to the restriction of w_α to Ω defines a bounded linear operator from $L^2(\Omega,\mathbb{C}^3)$ to $H_\alpha(\operatorname{curl},\Omega)$.

Proof.

(i) First, we remark that by the identities $\mathfrak{G}_{-\alpha}(x,y) = \mathfrak{G}_\alpha(y,x)$ and $\operatorname{curl}\operatorname{grad} = 0$ the integration kernel of $\mathcal{W}$ equals

$$\operatorname{curl}_x(\mathfrak{G}_\alpha(y,x)\, g(y)) = \operatorname{curl}_x(G_\alpha(y,x)\, \mathbb{I}_{3\times 3}\, g(y)) = \operatorname{curl}_x(G_\alpha(y,x)\, g(y))$$

and hence

$$w_\alpha(x) = \operatorname{curl} \int_\Omega G_\alpha(y,x)\, g(y)\, \mathrm{d}y. \tag{4.39}$$

We also remark that, even if the connected components of $\Gamma = \partial\Omega \cap \Pi$ are C^2-regular, Ω might still be just Lipschitz. Now, let $g \in C^{1,\gamma}(\overline{\Omega}, \mathbb{C}^3)$ with $0 < \gamma \leq 1$ and $a \in \mathbb{C}^3$ be a fixed vector. Analogous to the proof of Lemma 5.2 (a) in [45], with $x \notin \partial\Omega$ we obtain

$$
\begin{aligned}
a \cdot w_\alpha(x) &= \int_\Omega a \cdot \operatorname{curl}_x(G_\alpha(y,x)\, g(y))\, \mathrm{d}y \\
&= -\int_\Omega a \cdot (\operatorname{grad}_y G_\alpha(y,x) \times g(y))\, \mathrm{d}y \\
&= \int_\Omega g(y) \cdot \operatorname{curl}_y(G_\alpha(y,x)\, a)\, \mathrm{d}y \\
&= \int_\Omega G_\alpha(y,x)\, a \cdot \operatorname{curl} g(y)\, \mathrm{d}y + \int_{\partial\Omega} G_\alpha(y,x)\, a \cdot (g(y) \times \nu(y))\, \mathrm{d}s(y),
\end{aligned}
$$

where we used the identity $\operatorname{grad}_x G_\alpha(y,x) = -\operatorname{grad}_y G_\alpha(y,x)$ and, for the last equality, Theorem 2.9 (ii). Since this holds for all $a \in \mathbb{C}^3$, we conclude that

$$
w_\alpha(x) = \int_\Omega G_\alpha(y,x) \operatorname{curl} g(y)\, \mathrm{d}y + \int_{\partial\Omega} G_\alpha(y,x)\, (g(y) \times \nu(y))\, \mathrm{d}s(y).
$$

We recall that the unit normal vector ν exists almost everywhere on $\partial\Omega$. Under consideration of the decomposition (3.17) of the Green's function G_α, from the regularity of the standard volume potential ([17, Theorem 8.2]) and the jump relations for the single-layer potential for Lipschitz domains ([53, Theorem 6.11]) we conclude that w_α does not jump across $\partial\Omega \supseteq \Gamma$. By Proposition 3.3 (i) and the properties of the single-layer potential we have that $w_\alpha \in C^2_\alpha(\Pi \backslash \Gamma, \mathbb{C}^3) \cap C_\alpha(\Pi, \mathbb{C}^3)$ and $\Delta w_\alpha + k_0^2 w_\alpha = -\operatorname{curl} g$ in $\Pi \backslash \Gamma$, where the right-hand side is extended by zero into Ω^{ext}. In addition, the divergence of w_α vanishes in $\Pi \backslash \Gamma$, which proves (4.35). To show (4.37), we note the following. The standard volume potential over Ω is H^2-regular in a neighborhood of Ω ([17, Theorem 8.2]), hence the (weak) curl of this potential does not jump across $\partial\Omega$. Moreover, since the normal vector ν exists everywhere on Γ and $g|_\Gamma \times \nu$ is a tangential field on Γ, the integrand $\nu(x) \times \operatorname{curl}_x(G_\alpha(y,x)\, (g(y) \times \nu(y)))$ has the same type of singularity on Γ as the kernel of the double-layer potential, cf. the proof of Theorem 2.26 in [16]. Finally, the jump relations of the double-layer potential ([53, Theorem 6.11]) yield (4.37).

(ii) The asserted regularity of w_α follows from the decomposition (3.17) of the Green's function, the H^2-regularity of the standard volume potential (cf. [17, Theorem 8.2]), and the relation (4.39). Since under the conditions of (i) the potential w_α solves the variational equation (4.38), a denseness argument implies this also for $g \in L^2(\Omega, \mathbb{C}^3)$. Note that we do not need to make any further regularity assumption on Γ since the support of $g \in L^2(\Omega, \mathbb{C}^3)$ can have any regularity.

(iii) This is a consequence of the definition of w_α and part (ii). □

We remark that the operators $\mathcal{W}$ from (4.34) and $\mathcal{H}^*_{\Gamma_s}$ from (4.29) are related by

$$\mathcal{H}^*_{\Gamma_s} g = \mathcal{W}\big(\sqrt{|q|}\, g\big)\big|_{\Gamma_s}, \qquad g \in L^2(\Omega, \mathbb{C}^3).$$

By Proposition 4.3 (ii), a solution to the integro-differential equation

$$v_\alpha(x) = \operatorname{curl} \int_\Omega \mathfrak{G}_\alpha(y,x)\, \frac{q(y)}{\sqrt{|q(y)|}} \big(\hat{f}(y) + \sqrt{|q(y)|}\, \operatorname{curl} v_\alpha(y)\big)\, \mathrm{d}y \tag{4.40}$$

in Π is a radiating solution to (4.30). The equation (4.40) is called the α-quasi-periodic *electromagnetic Lippmann-Schwinger equation.* Since the only radiating solution to $\operatorname{curl}^2 \widetilde{v}_\alpha - k_0^2 \widetilde{v}_\alpha = 0$ in Π is $\widetilde{v}_\alpha \equiv 0$, the unique solution to (4.30) satisfies (4.40). Hence, the formulations (4.30) and (4.40) are equivalent, and we can write (4.40) for short as

$$v_\alpha(x) = \operatorname{curl} \int_\Omega \mathfrak{G}_\alpha(y,x)\, (\mathcal{T}\hat{f})(y)\, \sqrt{|q(y)|}\, \mathrm{d}y. \tag{4.41}$$

Now, in view of (4.29) and $\mathfrak{G}_{-\alpha}(x,y) = \mathfrak{G}_\alpha(y,x)$, (4.41) reveals the identity $\mathcal{H}^*_{\Gamma_s} \mathcal{T} \hat{f} = v_\alpha|_{\Gamma_s} = \mathcal{G}\hat{f}$. Combining this with (4.31), we arrive at the factorization

$$\mathcal{M} = \mathcal{H}^*_{\Gamma_s} \mathcal{T} \mathcal{H}_{\Gamma_s} \tag{4.42}$$

for the artificial near field operator. The connection between the variational formulation (4.30) and the integro-differential equation (4.40) is made precise in the following corollary to Proposition 4.3, cp. Theorem 2.3 in [43].

Corollary 4.4. *Under Assumptions 4.2 there hold:*

(i) If $v_\alpha \in H_{\alpha,loc}(\operatorname{curl}, \Pi)$ is a radiating solution to (4.30), then the restriction $v_\alpha|_\Omega \in H_\alpha(\operatorname{curl}, \Omega)$ solves the equation

$$\widetilde{v}_\alpha = \mathcal{W}\left(\frac{q}{\sqrt{|q|}} \hat{f} - q \operatorname{curl} \widetilde{v}_\alpha \right)\bigg|_\Omega . \tag{4.43}$$

(ii) If $v_\alpha \in H_\alpha(\operatorname{curl}, \Omega)$ solves (4.43), then it can be extended by the right-hand side of (4.40) to a radiating solution to (4.30).

In [34], HOHAGE devises a fast numerical solver for the electromagnetic Lippmann-Schwinger equation with a Hölder contrast $q \in C^{1,\gamma}(\mathbb{R}^3)$, but formulated for a bounded scatterer and for the total electric field rather than for the scattered magnetic field as done in (4.40). We have some hope that an efficient solver for (4.40) can be constructed in a similar fashion. In the last chapter of this thesis, we set up a solver for the considerably simpler acoustic Lippmann-Schwinger equation, which nevertheless shares many basic ideas with [34] as both methods are inspired by VAINIKKO'S primal article [68]. Moreover, we mention that KIRSCH investigates an equation related to, but more general than (4.40) in [43]. We close the subsection with the proof of some properties of the operator $\mathcal{H}_{\Gamma_s}$ and its adjoint, which become important later on. Subsection 4.2.3 complements this by a discussion of the inner operator $\mathcal{T}$ in the factorization (4.42).

Proposition 4.5.

(i) The operators $\mathcal{H}_{\Gamma_s}$ and $\mathcal{H}_{\Gamma_s}^*$ are compact.

(ii) The operator $\mathcal{H}_{\Gamma_s}$ is injective.

Proof.

(i) The operator $\mathcal{H}_{\Gamma_s}^* : L^2(\Omega,\mathbb{C}^3) \to L^2(\Gamma_s,\mathbb{C}^3)$ (see (4.29)) can be restated as

$$(\mathcal{H}_{\Gamma_s}^* g)(x) = \int_\Omega \operatorname{curl}_x(G_{-\alpha}(x,y)\,\mathbb{I}_{3\times 3})\,g(y)\,\sqrt{|q(y)|}\,\mathrm{d}y, \qquad x \in \Gamma_s.$$

This is a Hilbert-Schmidt integral operator with kernel in $L^2(\Gamma_s \times \Omega, \mathbb{C}^{3\times 3})$ and thus compact [59, Theorem 7.83]. The compactness of $\mathcal{H}_{\Gamma_s}$ is a direct implication [60, Theorem 4.19].

(ii) Let $\varphi \in \ker \mathcal{H}_{\Gamma_s}$. Since $q \neq 0$ a.e. in Ω, we have that the potential $h_\alpha \in H_{\alpha,\mathrm{loc}}(\operatorname{curl},\Pi)$ given by

$$h_\alpha(x) = \operatorname{curl} \int_{\Gamma_s} \overline{G_{-\alpha}(y,x)}\,\varphi(y)\,\mathrm{d}s(y), \qquad x \in \Pi,$$

vanishes in Ω, cp. (4.28). An analytic continuation argument shows that h_α vanishes in $\{x \in \Pi : m_- < x_3 < m_+\}$ with m_- and m_+ as specified on p. 11. By the jump relation $[\gamma_t(\operatorname{curl} h_\alpha)]_{\Gamma_+ \cup \Gamma_-} = 0$ (see p. 354 of [44]) we obtain $\gamma_{t,+}(\operatorname{curl} h_\alpha) = 0$ on $\Gamma_+ \cup \Gamma_-$, where $\gamma_{t,+}$ is the tangential trace operator for $R_+ \cup R_-$. Moreover, h_α satisfies the homogeneous second-order Maxwell's equation $\operatorname{curl}^2 h_\alpha - k_0^2 h_\alpha = 0$ in $R_+ \cup R_-$ and a radiation condition of the form

$$h_\alpha(x) = \sum_{z \in Z} \widetilde{h}_z^{\pm}(\varphi)\, e^{\mathrm{i}(\alpha_z \cdot x \mp \beta_z x_3)} \qquad \text{in } R_\pm, \tag{4.44}$$

with vectorial coefficients $\widetilde{h}_z^{\pm}(\varphi)$. By applying another 'curl' to the second-order Maxwell's equation and defining a function $\widetilde{h}_\alpha$ by the equation $\operatorname{curl} h_\alpha + \mathrm{i} k_0 \widetilde{h}_\alpha = 0$ in $R_+ \cup R_-$, for this $\widetilde{h}_\alpha$ we derive the problem

$$\left.\begin{array}{rl} \Delta \widetilde{h}_\alpha + k_0^2 \widetilde{h}_\alpha = 0 & \text{in } R_+ \cup R_- \\ \gamma_{D,+}(\operatorname{div} \widetilde{h}_\alpha) = 0 & \text{on } \Gamma_+ \cup \Gamma_- \\ \gamma_{t,+} \widetilde{h}_\alpha = 0 & \text{on } \Gamma_+ \cup \Gamma_- \end{array}\right\}. \tag{4.45}$$

This is an unusual, α-quasi-periodic *exterior electric boundary value problem*. The classical form of this problem for a bounded region is discussed in [16]. Due to unique solvability of (4.45), $\widetilde{h}_\alpha$ vanishes in $R_+ \cup R_-$. Now, we may use the complementary equation $\operatorname{curl} \widetilde{h}_\alpha - \mathrm{i} k_0 h_\alpha = 0$ to back-substitute $\widetilde{h}_\alpha$, since together with the one above it simply states the given homogeneous Maxwell's equation for h_α. We conclude that also h_α vanishes in $R_+ \cup R_-$. The jump relation $[\gamma_t h_\alpha]_{\Gamma_s} = \varphi$ finally implies $\varphi \equiv 0$. □

4.2.2 The interior transmission eigenvalue problem

As for the acoustic case, we want to expose shortly the *interior transmission eigenvalue problem* for the Maxwell's equations and its impact on the near field operator $\mathcal{M}$. We consider the problem of finding a solution (v_α, w_α) to the equation system

$$\left.\begin{array}{rrl} \operatorname{curl}\left(\frac{1}{\varepsilon_r} \operatorname{curl} v_\alpha\right) - k_0^2 v_\alpha = 0, & \operatorname{curl}^2 w_\alpha - k_0^2 w_\alpha = 0 & \text{in } \Omega \\ \gamma_t v_\alpha = \gamma_t w_\alpha, & \gamma_t\left(\frac{1}{\varepsilon_r} \operatorname{curl} v_\alpha\right) = \gamma_t(\operatorname{curl} w_\alpha) & \text{on } \Gamma \end{array}\right\}, \tag{4.46}$$

where γ_t denotes the tangential trace operator for Ω. For how to interpret (4.46) with regard to the regularity of the involved terms, we refer to the similar discussion of the variational formulation (4.11) for the direct problem.

Definition 4.6. The value k_0^2 is said to be an *interior transmission eigenvalue* with corresponding *eigenpair* $(v_\alpha, w_\alpha) \in H_\alpha(\operatorname{curl}, \Omega) \times H_\alpha(\operatorname{curl}, \Omega)$ if $(v_\alpha, w_\alpha) \neq$

$(0,0)$ and (v_α, w_α) satisfy $\gamma_t v_\alpha = \gamma_t w_\alpha$ on Γ,

$$\int_\Omega \left(\frac{1}{\varepsilon_r} \operatorname{curl} v_\alpha \cdot \operatorname{curl} \psi_{-\alpha} - k_0^2 v_\alpha \cdot \psi_{-\alpha} \right) \mathrm{d}x$$

$$= \int_\Omega (\operatorname{curl} w_\alpha \cdot \operatorname{curl} \psi_{-\alpha} - k_0^2 w_\alpha \cdot \psi_{-\alpha}) \,\mathrm{d}x \quad (4.47)$$

for all $\psi_{-\alpha} \in H_{-\alpha}(\operatorname{curl}, \Omega)$, and

$$\int_\Omega (\operatorname{curl} w_\alpha \cdot \operatorname{curl} \psi_{-\alpha} - k_0^2 w_\alpha \cdot \psi_{-\alpha}) \,\mathrm{d}x = 0 \qquad (4.48)$$

for all $\psi_{-\alpha} \in H_{-\alpha,(0)}(\operatorname{curl}, \Omega) = \{u_{-\alpha} \in H_{-\alpha}(\operatorname{curl}, \Omega) : \gamma_T u_{-\alpha} = 0 \text{ on } \Gamma\}$.

The formulation (4.47) is based on a separate application of the Green's identity (2.25) to the first both equations in (4.46). If the functions v_α and w_α of an eigenpair are sufficiently regular, then (v_α, w_α) is a variational solution to the system (4.46). In this case, the absence of any boundary integral in (4.47) accounts for the second coupling boundary condition in (4.46). We now give a first result on conditions under which interior transmission eigenvalues do *not* exist.

Proposition 4.7.

(i) If $\operatorname{Im} q > 0$ *a.e. in* Ω*, then* $k_0^2 > 0$ *is no interior transmission eigenvalue.*

(ii) Suppose that every connected component ω_l *of* Ω *can be decomposed into non-empty subdomains* $\Omega_j^{(l)}$ *with piecewise analytic boundaries* $\partial\Omega_j^{(l)}$*. Moreover, assume that the contrast* q *is analytic on each* $\Omega_j^{(l)}$ *and there is at least one subdomain* $\Omega_{j_0}^{(l)}$ *of each component* ω_l *such that* $\operatorname{Im} q(x) > 0$ *for all* $x \in \Omega_{j_0}^{(l)}$ *and* $\partial\Omega_{j_0}^{(l)} \cap \Gamma$ *has non-empty relative interior. Then* $k_0^2 > 0$ *is no interior transmission eigenvalue.*

Proof.

(i) We argue almost exactly as in the proof of Theorem 3.2 in [43]. Let $(v_\alpha, w_\alpha) \in H_\alpha(\operatorname{curl}, \Omega) \times H_\alpha(\operatorname{curl}, \Omega)$ solve (4.47) and (4.48) with $\gamma_t v_\alpha = \gamma_t w_\alpha$ on Γ. For $\psi_{-\alpha} = \overline{v_\alpha}$, from (4.47) we obtain

$$\int_\Omega \left(\frac{1}{\varepsilon_r} |\operatorname{curl} v_\alpha|^2 - k_0^2 |v_\alpha|^2 \right) \mathrm{d}x$$
$$= \int_\Omega (|\operatorname{curl} w_\alpha|^2 - k_0^2 |w_\alpha|^2) \,\mathrm{d}x + \int_\Omega (\operatorname{curl} w_\alpha \cdot \operatorname{curl} \overline{z_\alpha} - k_0^2 w_\alpha \cdot \overline{z_\alpha}) \,\mathrm{d}x$$
$$= \int_\Omega (|\operatorname{curl} w_\alpha|^2 - k_0^2 |w_\alpha|^2) \,\mathrm{d}x,$$

where $z_\alpha = v_\alpha - w_\alpha$ and the last equality holds by (4.48) due to $\gamma_t \overline{z_\alpha} = 0 \Rightarrow \gamma_T \overline{z_\alpha} = 0 \Rightarrow \overline{z_\alpha} \in H_{-\alpha,(0)}(\mathrm{curl}, \Omega)$. Taking the imaginary part, with $\operatorname{Im} q > 0 \Leftrightarrow \operatorname{Im}(1/\varepsilon_r) < 0$ a.e. we conclude that $\operatorname{curl} v_\alpha$ vanishes identically. Inserting this into (4.47) and using (4.48) shows that $\int_\Omega k_0^2 v_\alpha \cdot \psi_{-\alpha} \, \mathrm{d}x = 0$ for all $\psi_{-\alpha} \in H_{-\alpha,(0)}(\mathrm{curl}, \Omega)$ and thus $v_\alpha \equiv 0$ by $k_0^2 > 0$ and a denseness argument. Then, we see that w_α solves $\mathrm{curl}^2 w_\alpha - k_0^2 w_\alpha = 0$ in Ω (in the variational sense) together with $\gamma_t w_\alpha = 0$ and $\gamma_t(\operatorname{curl} w_\alpha) = 0$ on Γ. By the representation Theorem 4.1, we finally get that also $w_\alpha \equiv 0$. Thus, according to Definition 4.6, k_0^2 is no interior transmission eigenvalue.

(ii) We restrict the discussion first to a single connected component ω_l and let $\Omega_{j_0}^{(l)} \subseteq \omega_l$ be a subdomain as described in the statement. To beautify notation, we write $\Omega_{j_0}^{(l)}$ as Ω_0 for short. We repeat the argumentation of (i) up to the point where we can conclude that $\operatorname{curl} v_\alpha$ vanishes in Ω_0. Then, from the variational formulation of the first equation in (4.46), we obtain that $\int_{\Omega_0} k_0^2 v_\alpha \cdot \psi_{-\alpha} \, \mathrm{d}x = 0$ for all $\psi_{-\alpha} \in H_{-\alpha,(0)}(\mathrm{curl}, \Omega)$ with compact support in Ω_0. Thus, by $k_0^2 > 0$ and a denseness argument, v_α vanishes identically in Ω_0. Now, we define a function $\widetilde{v}_\alpha$ by the equation $\operatorname{curl} v_\alpha + \mathrm{i} k_0 \varepsilon_r \widetilde{v}_\alpha = 0$ in Ω_0, understood in the variational sense. We note that, since q is analytic in Ω_0, ε_r is analytic and has no roots in Ω_0, hence $\widetilde{v}_\alpha$ is well-defined. From the first equation in (4.46) we deduce the complementary equation $\operatorname{curl} \widetilde{v}_\alpha - \mathrm{i} k_0 v_\alpha = 0$ in Ω_0, both understood in the variational sense. With $v_\alpha = \widetilde{v}_\alpha = 0$ in Ω_0, we are now in a similar situation as in the proof of Theorem 7 in [63] and conclude that v_α and $\widetilde{v}_\alpha$ vanish in the whole connected component ω_l. Thus, w_α solves $\mathrm{curl}^2 w_\alpha - k_0^2 w_\alpha = 0$ in ω_l (in the variational sense) together with $\gamma_t w_\alpha = 0$ and $\gamma_t(\operatorname{curl} w_\alpha) = 0$ on $\partial \omega_l \cap \Gamma$. Similar to above, we define a function $\widetilde{w}_\alpha$ by $\operatorname{curl} w_\alpha + \mathrm{i} k_0 \widetilde{w}_\alpha = 0$ in ω_l and observe $\operatorname{curl} \widetilde{w}_\alpha - \mathrm{i} k_0 w_\alpha = 0$. Again, by the arguments in the proof of [63, Theorem 7], we find that w_α vanishes in ω_l. Argueing likewise for the other connected components yields $v_\alpha = w_\alpha \equiv 0$ in Ω, which finishes the proof. □

The next proposition points out the impact of this special transmission problem on the near field operator $\mathcal{M}$, cp. Theorem 4.4 (d) in [45].

Proposition 4.8. *Assume that k_0^2 is not an interior transmission eigenvalue. Then the operator $\mathcal{M} : L^2(\Gamma_s, \mathbb{C}^3) \to L^2(\Gamma_s, \mathbb{C}^3)$ is injective and has dense range in $L^2(\Gamma_s, \mathbb{C}^3)$.*

Proof. We recall that $\mathcal{M}$ refers to the artificial incident field

$$u^i_\alpha(x) = \int_{\Gamma_s} \overline{\mathfrak{G}_{-\alpha}(y,x)}\,\varphi(y)\,\mathrm{ds}(y), \qquad x \in \Pi\backslash\Gamma_s,$$

which generates the scattered field

$$u^s_\alpha(x) = \int_{\Gamma_s} u^s_{p,\alpha}(x,y)\,\varphi(y)\,\mathrm{ds}(y), \qquad x \in \Pi.$$

Now, let $\varphi \in \ker\mathcal{M}$. Then u^s_α has a vanishing near field $\mathcal{M}\varphi \equiv 0$ on Γ_s. The radiating behavior of u^s_α, an analytic continuation argument, and the fact that Ω' has no inclusions of the background medium imply that u^s_α vanishes identically in Ω^{ext}. Therefore, the functions $v_\alpha = u^i_\alpha + u^s_\alpha$ and $w_\alpha = u^i_\alpha$ lead to a solution pair for the interior transmission eigenvalue problem (4.47). From the assumption that k_0^2 is not an interior transmission eigenvalue, there follows $v_\alpha = w_\alpha = 0$ in Ω. By analytic continuation, u^i_α is then seen to vanish in $\{x \in \Pi : m_- < x_3 < m_+\}$. Using the jump relation $[\gamma_t(\mathrm{curl}^2 u^i_\alpha)]_{\Gamma_+\cup\Gamma_-} = 0$ (see Theorem 6.11 in [17]), we obtain $\gamma_{t,+}(\mathrm{curl}^2 u^i_\alpha) = k_0^2\,\gamma_{t,+}u^i_\alpha = 0$ on $\Gamma_+\cup\Gamma_-$. Here, $\gamma_{t,+}$ denotes the tangential trace operator for $R_+\cup R_-$. Similar to the proof of Proposition 4.5 (ii), we now formulate an α-quasi-periodic exterior electric boundary value problem for u^i_α and conclude that u^i_α vanishes in $R_+\cup R_-$. Finally, the jump relation $[\gamma_t(\mathrm{curl}\,u^i_\alpha)]_{\Gamma_s} = \varphi$ yields $\varphi \equiv 0$, showing that $\mathcal{M}$ is injective. The second part of the assertion can be proven following the idea of the proof of Proposition 3.7. □

4.2.3 The inner operator

In order to set up a Factorization Method for the electromagnetic inverse problem, we also need to inspect the inner operator $\mathcal{T}$ in the factorization (4.42), defined in (4.32). The following theorem establishes the same properties for $\mathcal{T}$ as Theorem 3.8 does for the inner operator T in the acoustic setting.

Theorem 4.9. *Let Assumptions 4.2 hold.*

(i) The operator $\mathcal{T}$ can be written in the form $\mathcal{T} = \widetilde{\mathcal{T}} + \mathcal{K}$ where $\mathcal{K} : L^2(\Omega,\mathbb{C}^3) \to L^2(\Omega,\mathbb{C}^3)$ is compact and, if q is real-valued and $q > 0$ a.e. in Ω, the operator $\mathrm{Re}\,\widetilde{\mathcal{T}}$ is coercive, precisely,

$$\mathrm{Re}\,\langle\widetilde{\mathcal{T}}\hat{f},\hat{f}\rangle_{L^2} \geq \|\hat{f}\|^2_{L^2} \tag{4.49}$$

holds for all $\hat{f} \in L^2(\Omega,\mathbb{C}^3)$.

(ii) *The operator* $\operatorname{Im}\mathcal{T}$ *is positive semi-definite, i.e.* $\operatorname{Im}\langle \mathcal{T}\hat{f}, \hat{f}\rangle_{L^2} \geq 0$ *for all* $\hat{f} \in L^2(\Omega, \mathbb{C}^3)$.

(iii) *Assume that there is a constant* $c_1 > 0$ *such that* $\operatorname{Im} q \geq c_1|q|$ *holds a.e. in* Ω. *Then* $\operatorname{Im}\mathcal{T}$ *is coercive with*

$$\operatorname{Im}\langle \mathcal{T}\hat{f}, \hat{f}\rangle_{L^2} \geq c_2 \|\hat{f}\|_{L^2}^2$$

for all $\hat{f} \in L^2(\Omega, \mathbb{C}^3)$ *and some constant* $c_2 > 0$.

(iv) *The operator* $\mathcal{T}$ *is injective.*

Proof. We make some preparations first and for these use an idea from the proof of Theorem 5.12 in [45]. We recall that $\mathcal{T} : L^2(\Omega, \mathbb{C}^3) \to L^2(\Omega, \mathbb{C}^3)$ is defined by

$$\mathcal{T}\hat{f} = \operatorname{sign}(q)\big(\hat{f} + \sqrt{|q|}\operatorname{curl} v_\alpha\big)\big|_\Omega$$

where $v_\alpha \in H_{\alpha,\mathrm{loc}}(\operatorname{curl}, \Pi)$ is a radiating variational solution to

$$\operatorname{curl}\left(\frac{1}{\varepsilon_r}\operatorname{curl} v_\alpha\right) - k_0^2 v_\alpha = \operatorname{curl}\left(\frac{q}{\sqrt{|q|}}\hat{f}\right) \qquad \text{in } \Pi.$$

This means that v_α solves, for all $\psi_{-\alpha} \in H_{-\alpha}(\operatorname{curl}, \Pi)$ with compact support in the x_3-dimension, the variational equation (4.30), which we repeat here for reference as

$$\int_\Pi \left(\frac{1}{\varepsilon_r}\operatorname{curl} v_\alpha \cdot \operatorname{curl} \psi_{-\alpha} - k_0^2 v_\alpha \cdot \psi_{-\alpha}\right) \mathrm{d}x = \int_\Omega \frac{q}{\sqrt{|q|}}\hat{f} \cdot \operatorname{curl} \psi_{-\alpha}\,\mathrm{d}x. \tag{4.50}$$

Now, we define $C_r = \{x \in \Pi : |x_3| < r\}$ for any $r > 0$ such that $\overline{\Omega} \cap \Pi \subset C_r$ and $\widetilde{\Pi} = C_{2r} \backslash \overline{C_r}$. Let further $\phi \in C^\infty(\Pi)$ be a mollifier with $\phi = 1$ in $\overline{C_r} \cap \Pi$ and $\phi = 0$ in $\Pi \backslash C_{2r}$. Then (4.50) states in particular that with $\psi_{-\alpha} = \phi\,\overline{v_\alpha} \in H_{-\alpha}(\operatorname{curl}, \Pi)$ there holds

$$\int_\Omega \frac{q}{\sqrt{|q|}}\hat{f} \cdot \operatorname{curl}\overline{v_\alpha}\,\mathrm{d}x = \int_{\overline{C_r}\cap\Pi} \left(\frac{1}{\varepsilon_r}|\operatorname{curl} v_\alpha|^2 - k_0^2 |v_\alpha|^2\right)\mathrm{d}x + \\ + \int_{\widetilde{\Pi}} \big(\operatorname{curl} v_\alpha \cdot \operatorname{curl}(\phi\,\overline{v_\alpha}) - k_0^2 \phi\,|v_\alpha|^2\big)\,\mathrm{d}x.$$

By an application of the Green's identity (2.24), we find that

$$\int_{\widetilde{\Pi}} \big(\operatorname{curl} v_\alpha \cdot \operatorname{curl}(\phi\,\overline{v_\alpha}) - \operatorname{curl}^2 v_\alpha \cdot (\phi\,\overline{v_\alpha})\big)\,\mathrm{d}x = -\int_{\partial\widetilde{\Pi}} (\nu \times \operatorname{curl} v_\alpha) \cdot (\phi\,\overline{v_\alpha})\,\mathrm{d}s$$

is equivalent to

$$\int_{\widetilde{\Pi}} \left(\operatorname{curl} v_\alpha \cdot \operatorname{curl}(\phi\, \overline{v_\alpha}) - k_0^2\, \phi\, |v_\alpha|^2\right) \mathrm{d}x = -\int_{\partial C_r \cap \Pi} (\operatorname{curl} v_\alpha \times \nu) \cdot \overline{v_\alpha}\, \mathrm{d}s$$

where ν denotes the exterior unit normal to $\partial C_r \cap \Pi$. Here, we have used that ϕ vanishes on $\partial C_{2r} \cap \Pi$. Thus, we obtain

$$\int_\Omega \frac{q}{\sqrt{|q|}} \hat{f} \cdot \operatorname{curl} \overline{v_\alpha}\, \mathrm{d}x = \int_{\overline{C_r} \cap \Pi} \left(\frac{1}{\varepsilon_r} |\operatorname{curl} v_\alpha|^2 - k_0^2\, |v_\alpha|^2\right) \mathrm{d}x - \\ - \int_{\partial C_r \cap \Pi} (\operatorname{curl} v_\alpha \times \nu) \cdot \overline{v_\alpha}\, \mathrm{d}s. \quad (4.51)$$

The left-hand side of this equation, and so the right-hand side, is actually independent of the parameter r. In a last preparatory step, we let r go to infinity and derive a convenient expression for

$$\lim_{r \to \infty} \int_{\partial C_r \cap \Pi} (\operatorname{curl} v_\alpha \times \nu) \cdot \overline{v_\alpha}\, \mathrm{d}s.$$

Since v_α radiates, it has the Rayleigh expansion

$$v_\alpha(x) = \sum_{z \in Z} v_z^\pm\, e^{\mathrm{i}(\alpha_z \cdot x \pm \beta_z x_3)} \qquad \text{in } R_\pm \quad (4.52)$$

with the Rayleigh coefficient vectors

$$v_z^\pm = \mathrm{i} \begin{pmatrix} \alpha_{z,1} \\ \alpha_{z,2} \\ \pm\beta_z \end{pmatrix} \times \widetilde{v}_z^\pm.$$

Using this, a straightforward computation leads to

$$\operatorname{curl} v_\alpha(x) = \sum_{z \in Z} \left(k_0^2\, \widetilde{v}_z^\pm - \begin{pmatrix} \alpha_{z,1} \\ \alpha_{z,2} \\ \pm\beta_z \end{pmatrix} \left[\begin{pmatrix} \alpha_{z,1} \\ \alpha_{z,2} \\ \pm\beta_z \end{pmatrix} \cdot \widetilde{v}_z^\pm \right] \right) E_z^\pm(x) \qquad \text{in } R_\pm$$

where $E_z^\pm(x) = e^{\mathrm{i}(\alpha_z \cdot x \pm \beta_z x_3)}$. For $x \in \partial C_r \cap \Pi$ there hold $|x_3| = r$ and either $\nu(x) = e_3$ or $\nu(x) = -e_3$. In (4.52), the partial sum over all terms corresponding to $z \in Z$ with $\beta_z \in \mathrm{i}\,\mathbb{R}^+$ vanishes for $|x_3| \to \infty$. Moreover, $e^{\mathrm{i}(\alpha_z \cdot x)}$ with $z \in Z$ as functions of $(x_1, x_2) \in (-\pi, \pi)^2$ are orthogonal to each other. A technical calculation reveals that

$$\lim_{r \to \infty} \int_{\partial C_r \cap \Pi} (\operatorname{curl} v_\alpha \times \nu) \cdot \overline{v_\alpha}\, \mathrm{d}s = \mathrm{i}\, 4\pi^2 \sum_{\beta_z \in \mathbb{R}^+} \beta_z |v_z^\pm|^2. \quad (4.53)$$

Combining (4.53) with (4.51), we finally arrive at

$$\int_\Omega \frac{q}{\sqrt{|q|}} \hat{f} \cdot \operatorname{curl} \overline{v_\alpha} \, \mathrm{d}x = \int_\Pi \left(\frac{1}{\varepsilon_r} |\operatorname{curl} v_\alpha|^2 - k_0^2 |v_\alpha|^2 \right) \mathrm{d}x - \mathrm{i} 4\pi^2 \sum_{\beta_z \in \mathbb{R}^+} \beta_z |v_z^\pm|^2. \tag{4.54}$$

We will exploit this relation in the proof of parts (i)–(iii).

(i) We note first that the operator $\mathcal{T}$ depends on the wave number k_0 via v_α solving (4.50). We make this dependence explicit here by writing $\mathcal{T}_{k_0}$ and v_{α,k_0}, respectively. Now, we decompose $\mathcal{T}_{k_0}$ as

$$\mathcal{T}_{k_0} = \mathcal{T}_\mathrm{i} + (\mathcal{T}_{k_0} - \mathcal{T}_\mathrm{i})$$

and show that $\widetilde{\mathcal{T}} = \mathcal{T}_\mathrm{i}$ and $\mathcal{K} = \mathcal{T}_{k_0} - \mathcal{T}_\mathrm{i}$ have the asserted properties. We consider the hypothetical case $k_0 = \mathrm{i}$ here only to define the auxiliary operator $\mathcal{T}_\mathrm{i}$; apart from that, we assume $k_0 = \omega\sqrt{\varepsilon_0 \mu_0} \in \mathbb{R}^+$. The operator $\mathcal{T}_\mathrm{i} : L^2(\Omega, \mathbb{C}^3) \to L^2(\Omega, \mathbb{C}^3)$ is given by

$$\mathcal{T}_\mathrm{i} \hat{f} = \operatorname{sign}(q) \big(\hat{f} + \sqrt{|q|} \operatorname{curl} v_{\alpha,\mathrm{i}} \big) \big|_\Omega,$$

where $v_{\alpha,\mathrm{i}} \in H_{\alpha,\mathrm{loc}}(\operatorname{curl}, \Pi)$ radiates and satisfies

$$\int_\Pi \left(\frac{1}{\varepsilon_r} \operatorname{curl} v_{\alpha,\mathrm{i}} \cdot \operatorname{curl} \psi_{-\alpha} + v_{\alpha,\mathrm{i}} \cdot \psi_{-\alpha} \right) \mathrm{d}x = \int_\Omega \frac{q}{\sqrt{|q|}} \hat{f} \cdot \operatorname{curl} \psi_{-\alpha} \, \mathrm{d}x \tag{4.55}$$

for all $\psi_{-\alpha} \in H_{-\alpha}(\operatorname{curl}, \Pi)$ with compact support in the x_3-dimension. This operator is well-defined by the following argumentation, cf. the proof of [43, Lemma 2.4 (a)]. The equivalence between the variational formulation (4.50) for a radiating v_{α,k_0} and the integro-differential equation (4.40), stated by Corollary 4.4, also holds for $k_0 = \mathrm{i}$, cp. the proof of [43, Theorem 2.3]. We denote the Green's function G_α and the Green's tensor $\mathfrak{G}_\alpha$ more precisely here by G_{α,k_0} and $\mathfrak{G}_{\alpha,k_0}$, respectively. The definition of $\mathfrak{G}_{\alpha,\mathrm{i}}$ shows that every solution $v_{\alpha,\mathrm{i}}$ to (4.40) lies in fact in $H_\alpha(\operatorname{curl}, \Pi)$. In consequence, the variational equation (4.55) holds for all $\psi_{-\alpha} \in H_{-\alpha}(\operatorname{curl}, \Pi)$. Finally, the theorem of Lax-Milgram yields the existence of a unique solution to (4.55) and the boundedness of the operator from $L^2(\Omega, \mathbb{C}^3)$ to $H_\alpha(\operatorname{curl}, \Omega)$ which maps $\hat{f}$ to $v_{\alpha,\mathrm{i}}|_\Omega$. Thus, $\mathcal{T}_\mathrm{i} : L^2(\Omega, \mathbb{C}^3) \to L^2(\Omega, \mathbb{C}^3)$ is well-defined. Using (4.54), we find that

$$\operatorname{Re}\left(\int_\Omega \frac{q}{\sqrt{|q|}} \hat{f} \cdot \operatorname{curl} \overline{v_{\alpha,\mathrm{i}}} \, \mathrm{d}x \right) = \int_\Pi \left(\operatorname{Re}\left(\frac{1}{\varepsilon_r} \right) |\operatorname{curl} v_{\alpha,\mathrm{i}}|^2 + |v_{\alpha,\mathrm{i}}|^2 \right) \mathrm{d}x \geq 0,$$

where $\operatorname{Re}\varepsilon_r^{-1} = |\varepsilon_r|^{-2}\operatorname{Re}\varepsilon_r \geq 0$ by Assumptions 4.2. For real-valued contrast q, there holds

$$\operatorname{Re}\left(\int_\Omega \frac{q}{\sqrt{|q|}} \operatorname{curl} v_{\alpha,\mathrm{i}} \cdot \overline{\hat{f}}\,\mathrm{d}x\right) = \operatorname{Re}\left(\int_\Omega \frac{q}{\sqrt{|q|}} \hat{f} \cdot \operatorname{curl} \overline{v_{\alpha,\mathrm{i}}}\,\mathrm{d}x\right) \geq 0.$$

Hence, under $q > 0$ in Ω, which implies $\operatorname{sign}(q) \equiv 1$, we obtain

$$\operatorname{Re}\langle \mathcal{T}_\mathrm{i}\hat{f}, \hat{f}\rangle_{L^2} = \int_\Omega |\hat{f}|^2\,\mathrm{d}x + \operatorname{Re}\left(\int_\Omega \frac{q}{\sqrt{|q|}} \operatorname{curl} v_{\alpha,\mathrm{i}} \cdot \overline{\hat{f}}\,\mathrm{d}x\right) \geq \|\hat{f}\|_{L^2}^2,$$

which shows the coercivity of $\operatorname{Re}\widetilde{\mathcal{T}}$ for $\widetilde{\mathcal{T}} = \mathcal{T}_\mathrm{i}$. It remains to prove the compactness of the operator $\mathcal{K} = \mathcal{T}_{k_0} - \mathcal{T}_\mathrm{i}$, which reads

$$(\mathcal{T}_{k_0} - \mathcal{T}_\mathrm{i})\hat{f} = \frac{q}{\sqrt{|q|}} \operatorname{curl}(v_{\alpha,k_0} - v_{\alpha,\mathrm{i}})\big|_\Omega.$$

By the integro-differential equation (4.40) and the remarks in the proof of Proposition 4.3 (i), we can write $(v_{\alpha,k_0} - v_{\alpha,\mathrm{i}})|_\Omega$ as

$$\begin{aligned}
&(v_{\alpha,k_0} - v_{\alpha,\mathrm{i}})(x)\\
&= \operatorname{curl}\int_\Omega G_{\alpha,k_0}(y,x)\,q(y)\operatorname{curl}(v_{\alpha,k_0} - v_{\alpha,\mathrm{i}})(y)\,\mathrm{d}y +\\
&\quad + \operatorname{curl}\int_\Omega (G_{\alpha,k_0} - G_{\alpha,\mathrm{i}})(y,x)\,\frac{q(y)}{\sqrt{|q(y)|}}\big(\hat{f}(y) + \sqrt{|q(y)|}\operatorname{curl} v_{\alpha,\mathrm{i}}(y)\big)\,\mathrm{d}y
\end{aligned}$$

with $x \in \Omega$. Equivalently, $(v_{\alpha,k_0} - v_{\alpha,\mathrm{i}})|_\Omega$ satisfies the equation

$$(\mathrm{id} - \mathcal{B}_{k_0})(v_{\alpha,k_0} - v_{\alpha,\mathrm{i}})|_\Omega = \mathcal{C}_{k_0}\hat{f}$$

where the operators $\mathcal{B}_{k_0}$ and $\mathcal{C}_{k_0}$ are given by

$$(\mathcal{B}_{k_0}\widetilde{v}_\alpha)(x) = \operatorname{curl}\int_\Omega G_{\alpha,k_0}(y,x)\,q(y)\operatorname{curl}\widetilde{v}_\alpha(y)\,\mathrm{d}y, \qquad \widetilde{v}_\alpha \in H_\alpha(\operatorname{curl},\Omega),$$

and

$$\begin{aligned}
(\mathcal{C}_{k_0}\hat{f})(x) = \operatorname{curl}\int_\Omega (G_{\alpha,k_0} - G_{\alpha,\mathrm{i}})(y,x)\,\frac{q(y)}{\sqrt{|q(y)|}}\cdot&\\
\cdot\big(\hat{f}(y) + \sqrt{|q(y)|}\operatorname{curl} v_{\alpha,\mathrm{i}}(y)\big)\,\mathrm{d}y&
\end{aligned}$$

with $x \in \Omega$. By similar considerations as in the proofs of [43, Lemma 2.2] and [44, Lemma 3.1], together with the decomposition $G_{\alpha,k}(y,x) =$

$\Phi_k(x,y)+\Psi_k(x-y)$ in $(\Pi\times\Pi)\backslash\{(x,x) : x\in\Pi\}$ with the fundamental solution Φ_k to the Helmholtz equation in $\mathbb{R}^3$ and a smooth function Ψ_k on $2\cdot\Pi$ for arbitrary $k\in\mathbb{C}\backslash\{0\}$ with $\operatorname{Re}k\geq 0$ and $\operatorname{Im}k\geq 0$, we find that $\mathcal{B}_{k_0}$ is bounded from $H_\alpha(\operatorname{curl},\Omega)$ to $H_\alpha(\operatorname{curl},\Omega)$ and $\mathcal{C}_{k_0}$ is bounded from $L^2(\Omega,\mathbb{C}^3)$ to $H_\alpha(\operatorname{curl},\Omega)$. Moreover, using arguments analog to those in the proof of [43, Lemma 2.4 (b)], it can be shown that $\mathcal{C}_{k_0}$ is compact and $\mathrm{id}-\mathcal{B}_{k_0}$ is the sum of a boundedly invertible operator and a compact operator, i.e. $\mathrm{id}-\mathcal{B}_{k_0}$ is a Fredholm operator with index zero. Now, we note that the restriction of the integro-differential equation (4.40) to Ω can be written as

$$(\mathrm{id}-\mathcal{B}_{k_0})v_{\alpha,k_0}|_\Omega=\operatorname{curl}\int_\Omega G_{\alpha,k_0}(y,\cdot)\,\frac{q(y)}{\sqrt{|q(y)|}}\hat{f}(y)\,\mathrm{d}y. \tag{4.56}$$

Since we assume that the direct problem defined on p. 53 is uniquely solvable (cf. Assumptions 4.2) and the variational problem (4.50) as well as the integro-differential equations (4.40) and (4.56) are uniquely solvable alike, the operator $\mathrm{id}-\mathcal{B}_{k_0}$ is also injective. Hence, this operator is boundedly invertible in $H_\alpha(\operatorname{curl},\Omega)$, cf. Corollary 1.20 in [16]. The observation that

$$\mathcal{K}=\mathcal{T}_{k_0}-\mathcal{T}_\mathrm{i}=\frac{q}{\sqrt{|q|}}\operatorname{curl}(\mathrm{id}-\mathcal{B}_{k_0})^{-1}\,\mathcal{C}_{k_0}$$

finally reveals the compactness of $\mathcal{K}: L^2(\Omega,\mathbb{C}^3)\to L^2(\Omega,\mathbb{C}^3)$. The proof is finished.

(ii) We define $\hat{g}\in L^2(\Omega,\mathbb{C}^3)$ by $\hat{g}=(\hat{f}+\sqrt{|q|}\operatorname{curl}v_\alpha)\big|_\Omega$, so that

$$\langle\mathcal{T}\hat{f},\hat{f}\rangle_{L^2}=\int_\Omega\operatorname{sign}(q)\,|\hat{g}|^2\,\mathrm{d}x-\int_\Omega\frac{q}{\sqrt{|q|}}\hat{g}\cdot\operatorname{curl}\overline{v_\alpha}\,\mathrm{d}x. \tag{4.57}$$

For the second term on the right-hand side, we calculate

$$\begin{aligned}\int_\Omega\frac{q}{\sqrt{|q|}}\hat{g}\cdot\operatorname{curl}\overline{v_\alpha}\,\mathrm{d}x=\int_\Omega\left(|\operatorname{curl}v_\alpha|^2-\frac{1}{\varepsilon_r}|\operatorname{curl}v_\alpha|^2\right)\mathrm{d}x+\\+\int_\Omega\frac{q}{\sqrt{|q|}}\hat{f}\cdot\operatorname{curl}\overline{v_\alpha}\,\mathrm{d}x,\end{aligned}$$

and, plugging in the relation (4.54), we get

$$\int_\Omega\frac{q}{\sqrt{|q|}}\hat{g}\cdot\operatorname{curl}\overline{v_\alpha}\,\mathrm{d}x=\int_\Pi(|\operatorname{curl}v_\alpha|^2-k_0^2\,|v_\alpha|^2)\,\mathrm{d}x-\mathrm{i}\,4\pi^2\sum_{\beta_z\in\mathbb{R}^+}\beta_z|v_z^\pm|^2.$$

Using this expression in (4.57), we then obtain

$$\operatorname{Im}\langle \mathcal{T}\hat{f},\hat{f}\rangle_{L^2} = \int_\Omega \frac{\operatorname{Im} q}{|q|}\,|\hat{g}|^2\,\mathrm{d}x + 4\pi^2 \sum_{\beta_z\in\mathbb{R}^+} \beta_z |v_z^\pm|^2.$$

Since $\operatorname{Im} q \geq 0$ by Assumptions 4.2, the assertion is shown.

(iii) This follows from part (ii) and an argumentation similar to that in the proof of Theorem 5.12 (c) in [45].

(iv) Let $\hat{f} \in \ker \mathcal{T}$. Then (4.33) turns into the homogeneous Maxwell's equation $\operatorname{curl}^2 v_\alpha - k_0^2 v_\alpha = 0$ in Π. According to the discussion at the end of Subsection 4.1.1, for each component of v_α we can now argue as in Theorem 3.8 (iv) and conclude that v_α vanishes identically in Π. Then from

$$\begin{aligned} &\mathcal{T}\hat{f} = \operatorname{sign}(q)\big(\hat{f} + \sqrt{|q|}\,\operatorname{curl} v_\alpha\big)\big|_\Omega = 0 \\ \Longleftrightarrow\quad &\big(\hat{f} + \sqrt{|q|}\,\operatorname{curl} v_\alpha\big)\big|_\Omega = 0 \end{aligned}$$

there follows $\hat{f} \equiv 0$, hence $\mathcal{T}$ is injective. □

Chapter 5

Approximation of the near field operators

In Subsections 3.1.3 and 4.1.3 we introduced the acoustic and electromagnetic near field operators M and $\mathcal{M}$, respectively. These were the starting points in our construction of Factorization Methods for the acoustic and electromagnetic inverse problems and have been factorized and discussed subsequently. However, M and $\mathcal{M}$ are not the physically relevant near field operators and have been defined as auxiliary operators to meet a basic requirement of the Factorization Method. In this chapter, we prove that M and $\mathcal{M}$ can be approximated (arbitrarily well in theory) by means of only the physical near field operators $\widetilde{M}$ and $\widetilde{\mathcal{M}}$, see (3.23) and (4.26). These operators can be computed (approximately) from data known or measured in practice. Precisely, we will not require any knowledge which cannot be acquired from the data. Afterwards, in Chapter 6, we show that the functional analytic result which underlies the Factorization Method can be applied to M and $\mathcal{M}$, as opposed to the situation for the physical operators $\widetilde{M}$ and $\widetilde{\mathcal{M}}$.

5.1 The acoustic case

To begin, we examine the mapping properties of the operators involved in the acoustic problem. The operator H'_{Γ_s} defined by

$$(H'_{\Gamma_s}\varphi)(x) = \int_{\Gamma_s} \overline{G_{-\alpha}(y,x)}\, \varphi(y)\, \mathrm{ds}(y), \qquad x \in \Omega,$$

is an element of $\mathcal{L}(L^2(\Gamma_s), H^1_\alpha(\Omega))$ and related to H_{Γ_s} by $H_{\Gamma_s}\varphi = \sqrt{|q|}\, H'_{\Gamma_s}\varphi$, cp. (3.29). Having this, the preliminary factorization (3.32) states that

$$M\varphi = G\sqrt{|q|}\, H'_{\Gamma_s}\varphi \tag{5.1}$$

holds for all $\varphi \in L^2(\Gamma_s)$. Analogously, the physical near field operator $\widetilde{M}$ obeys

$$\widetilde{M}\phi = G\sqrt{|q|}\, \widetilde{H}'_{\Gamma_i}\phi \tag{5.2}$$

for all $\phi \in L^2(\Gamma_i)$ where $\widetilde{H}'_{\Gamma_i} \in \mathcal{L}(L^2(\Gamma_i), H^1_\alpha(\Omega))$ is given by

$$(\widetilde{H}'_{\Gamma_i}\phi)(x) = \int_{\Gamma_i} G_\alpha(y,x)\,\phi(y)\,\mathrm{d}s(y), \qquad x \in \Omega.$$

In (5.1) and (5.2), G is the solution operator defined on p. 41. Now, let $R = \{x \in \Pi : |x_3| < r\}$ with $r > 0$ chosen such that $\overline{\Omega} \cap \Pi \subset R$ and $\overline{R} \cap \Gamma_i = \overline{R} \cap \Gamma_s = \emptyset$. Our main tool in the approximation procedure for M are the operators $\widetilde{V}_{\Gamma_i} : L^2(\Gamma_i) \to H^{1/2}_\alpha(\partial R)$ and $V_{\Gamma_s} : L^2(\Gamma_s) \to H^{1/2}_\alpha(\partial R)$ defined as

$$\widetilde{V}_{\Gamma_i}\phi = \int_{\Gamma_i} G_\alpha(y,\cdot)\,\phi(y)\,\mathrm{d}s(y) \qquad \text{and} \qquad V_{\Gamma_s}\varphi = \int_{\Gamma_s} \overline{G_{-\alpha}(y,\cdot)}\,\varphi(y)\,\mathrm{d}s(y).$$

In the next lemma, we collect some important facts about $\widetilde{V}_{\Gamma_i}$ and V_{Γ_s}.

Lemma 5.1.

(i) The operators $\widetilde{V}_{\Gamma_i}$ and V_{Γ_s} are compact.

(ii) If k_0^2 is not a Dirichlet eigenvalue of $-\Delta$ in R, i.e. if the problem

$$\left.\begin{aligned} \Delta v_\alpha + k_0^2 v_\alpha &= 0 \quad && \text{in } R \\ \gamma_D v_\alpha &= 0 \quad && \text{on } \partial R \cap \Pi \end{aligned}\right\} \tag{5.3}$$

has only the trivial solution, then $\widetilde{V}_{\Gamma_i}$ and V_{Γ_s} are injective and their ranges are dense in $H^{1/2}_\alpha(\partial R)$. We note that the subscript α prescribes an α-quasi-periodic boundary condition on $\partial R \cap \partial\Pi$.

(iii) Assume that k_0^2 is not a Dirichlet eigenvalue of $-\Delta$ in R, such that $\widetilde{V}_{\Gamma_i}$ is injective. Let $\{\widetilde{P}_\delta\}_{\delta>0}$ be a family of operators $\widetilde{P}_\delta : H^{1/2}_\alpha(\partial R) \to L^2(\Gamma_i)$ which forms a regularization for the (unbounded) inverse of $\widetilde{V}_{\Gamma_i}$ on $\mathcal{R}(\widetilde{V}_{\Gamma_i})$, cf. [25]. Then with $P_\delta = \widetilde{P}_\delta V_{\Gamma_s} : L^2(\Gamma_s) \to L^2(\Gamma_i)$ for $\delta > 0$ there holds the pointwise convergence

$$\widetilde{V}_{\Gamma_i} P_\delta \varphi \to V_{\Gamma_s}\varphi \quad \text{in } H^{1/2}_\alpha(\partial R) \quad \text{for } \delta \to 0 \tag{5.4}$$

for every $\varphi \in L^2(\Gamma_s)$.

Proof.

(i) The compactness of $\widetilde{V}_{\Gamma_i}$ and V_{Γ_s} follows directly from the smoothness of $G_\alpha(y,x)$ for all $y \in \Gamma_i \cup \Gamma_s$ and all x in an open proper superset of R which does not intersect $\Gamma_i \cup \Gamma_s$. Such set exists due to the conditions on R.

(ii) The proof scheme for this part is adopted from the proof of Lemma 5.1 in [44]. Let $g \in \ker V_{\Gamma_s}$. Then, for the potential

$$v_\alpha(x) = \int_{\Gamma_s} \overline{G_{-\alpha}(y,x)}\, g(y)\, \mathrm{d}s(y), \qquad x \in \Pi,$$

there holds $v_\alpha|_{\partial R \cap \Pi} = (V_{\Gamma_s} g)|_{\partial R \cap \Pi} = 0$. Note that the restriction of v_α to $\partial R \cap \Pi$ is well-defined here. Due to the properties of the single-layer potential ([53, Theorem 6.11]), v_α lies in $H^1_{\alpha,\mathrm{loc}}(\Pi)$ and satisfies the jump relations

$$[\gamma_D v_\alpha]_{\Gamma_+ \cup \Gamma_-} = 0 \qquad \text{and} \qquad [\gamma_N v_\alpha]_{\Gamma_s} = -g. \tag{5.5}$$

Moreover, v_α solves the Helmholtz equation in $\Pi \backslash \Gamma_s$ and has an expansion of the form (3.26) in $R_\pm$. Hence, $v_\alpha|_R$ is a solution to the interior Dirichlet problem (5.3). From the assumption on k_0^2, we conclude that v_α vanishes in R and, by an analytic continuation argument, even in $\{x \in \Pi : m_- < x_3 < m_+\}$. Then, the first jump relation in (5.5) implies $\gamma_{D,+} v_\alpha = 0$ on $\Gamma_+ \cup \Gamma_-$, where $\gamma_{D,+}$ is the trace operator for $R_+ \cup R_-$. We are now in the same situation as in the proof of Proposition 3.5 (ii) and obtain that v_α vanishes in $R_+ \cup R_-$. Finally, the second jump relation yields $g \equiv 0$, hence V_{Γ_s} is injective. The argumentation for $\widetilde{V}_{\Gamma_i}$ is essentially the same. To prove the denseness of $\mathcal{R}(V_{\Gamma_s})$ in $H^{1/2}_\alpha(\partial R)$, we show that also the adjoint $V^*_{\Gamma_s} : H^{-1/2}_\alpha(\partial R) \to L^2(\Gamma_s)$ is injective. Again, the procedure for $\widetilde{V}^*_{\Gamma_i}$ is similar. The adjoint of V_{Γ_s} is easily found to read $(V^*_{\Gamma_s}\varphi)(x) = \int_{\partial R} G_\alpha(y,x)\,\varphi(y)\,\mathrm{d}s(y)$ for $\varphi \in H^{-1/2}_\alpha(\partial R)$ and $x \in \Gamma_s$. Let now $\varphi \in \ker V^*_{\Gamma_s}$. We define the potential $w_\alpha \in H^1_{\alpha,\mathrm{loc}}(2 \cdot \Pi)$ by

$$w_\alpha(x) = \int_{\partial R} G_\alpha(y,x)\,\varphi(y)\,\mathrm{d}s(y), \qquad x \in 2 \cdot \Pi,$$

such that $w_\alpha|_{\Gamma_s} = V^*_{\Gamma_s}\varphi = 0$. We use the extended domain $2 \cdot \Pi$ here since we will have to consider the behavior of w_α on both sides of the boundary $\partial R \not\subset \Pi$. Again by Theorem 6.11 in [53], there hold the jump relations

$$[\gamma_D w_\alpha]_{\partial R} = 0 \qquad \text{and} \qquad [\gamma_N w_\alpha]_{\partial R} = -\varphi. \tag{5.6}$$

The geometric setting is illustrated in Figure 5.1. We note first that, by the analyticity of $w_\alpha((\cdot,\cdot,x_3))$ in $\Gamma_+ \cup \Gamma_-$, $w_\alpha|_{\Gamma_s} = 0$ implies $w_\alpha|_{\Gamma_+ \cup \Gamma_-} = 0$. Moreover, w_α solves the Helmholtz equation in $\Pi \backslash \partial R$ and has a Rayleigh expansion in $\widetilde{R} = \{x \in 2 \cdot \Pi : |x_3| > r\}$. From the uniqueness of a radiating solution to the exterior Dirichlet problem in $\{x \in 2 \cdot \Pi : x_3 < m_- \vee x_3 > m_+\}$,

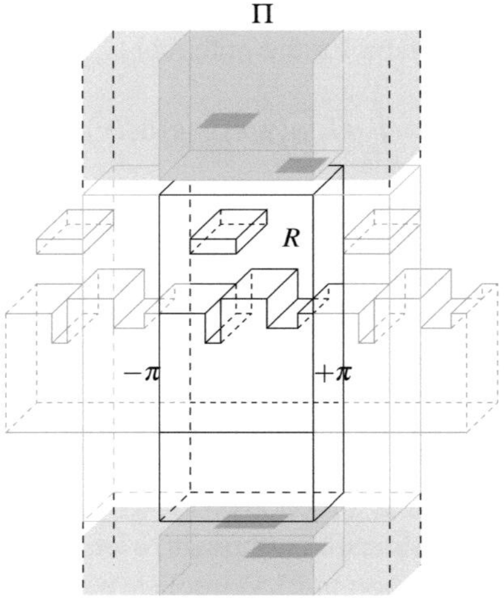

Figure 5.1: Extended domain

we conclude again by an analytic continuation argument that w_α vanishes in all of $\widetilde{R}$. As a consequence of the continuity of w_α across ∂R (in the sense of traces), we have $\gamma_{D,-} w_\alpha = 0$ on $\partial R \cap \Pi$, where $\gamma_{D,-}$ is the trace operator for R. Hence, $w_\alpha|_R$ solves the interior Dirichlet problem (5.3). Under the assumption on k_0^2, there follows $w_\alpha = 0$ in R. Exploiting the α-quasi-periodicity of w_α and the second jump relation in (5.6), we finally obtain $\varphi \equiv 0$. This completes the proof.

(iii) The pointwise convergence (5.4) follows from standard regularization theory, cf., e.g., [25]. We note that since by (ii) the operator $\widetilde{V}_{\Gamma_i} : L^2(\Gamma_i) \to H_\alpha^{1/2}(\partial R)$ is injective and has a dense range, its (unbounded) inverse on $\mathcal{R}(\widetilde{V}_{\Gamma_i})$ coincides with its generalized inverse $\widetilde{V}_{\Gamma_i}^\dagger : \mathcal{R}(\widetilde{V}_{\Gamma_i}) \oplus \mathcal{R}(\widetilde{V}_{\Gamma_i})^\perp \to L^2(\Gamma_i)$. In addition, we remark that convergence in the operator norm does not hold, i.e. $\|\widetilde{V}_{\Gamma_i} P_\delta - V_{\Gamma_s}\| \nrightarrow 0$. □

From now on, we assume that k_0^2 is not a Dirichlet eigenvalue of $-\Delta$ in R. Then the α-quasi-periodic interior Dirichlet problem

$$\left.\begin{aligned} \Delta v_\alpha + k_0^2 v_\alpha &= 0 \qquad \text{in } R \\ \gamma_D v_\alpha &= f_\alpha \qquad \text{on } \partial R \cap \Pi \end{aligned}\right\} \tag{5.7}$$

is uniquely solvable for every $f_\alpha \in C_\alpha(\partial R)$, cp. Theorem 3.24 in [16] for the classical interior Dirichlet problem for a bounded medium with C^2-boundary. Since with $\varphi \in L^2(\Gamma_s)$ both α-quasi-periodic potentials

$$x \longmapsto \int_{\Gamma_i} G_\alpha(y,x)\,(P_\delta\varphi)(y)\,\mathrm{ds}(y) \qquad \text{and} \qquad x \longmapsto \int_{\Gamma_s} \overline{G_{-\alpha}(y,x)}\,\varphi(y)\,\mathrm{ds}(y)$$

solve the Helmholtz equation in R, the pointwise convergence (5.4) and the continuous dependence of the solution to the problem (5.7) from its boundary data on ∂R yields that $\widetilde{H}'_{\Gamma_i} P_\delta \varphi$ converges to $H'_{\Gamma_s}\varphi$ in $H^1_\alpha(\Omega)$ as δ goes to zero. Thus, we obtain the pointwise convergence

$$\widetilde{M}P_\delta\varphi = G\sqrt{|q|}\,\widetilde{H}'_{\Gamma_i}P_\delta\varphi \to G\sqrt{|q|}\,H'_{\Gamma_s}\varphi = M\varphi \qquad \text{for } \delta \to 0 \tag{5.8}$$

for every $\varphi \in L^2(\Gamma_s)$. In view of (5.8), we can, in principle, evaluate the artificial near field operator M at any given instance φ with a prescribed accuracy, since the physical operator $\widetilde{M}$ can be constructed from measurements of the scattered field and the operator P_δ can be computed for any $\delta > 0$, provided that a set R with the above properties is known, see p. 78. (This requirement does not cause extra trouble as choosing a proper incidence surface Γ_i is essentially the same.) However, our final objective is the characterization of the range of an operator which depends on the exact operator M. It is an open question whether the pointwise convergence (5.8) is sufficient and convenient to accomplish this by means of the approximants $\widetilde{M}P_\delta$. For the tool we refer to later on, it is crucial that $\widetilde{M}P_\delta$ converges to M even in the operator norm. Although our proof of this fact has a different background, it is strongly inspired by the proof of Proposition III.12 in [49]. Beforehand, we prove an auxiliary result, following the next definition.

Definition 5.2. Let $L \subset \mathbb{R}^3$ be a bounded Lipschitz set. We define the operator $SL_{\partial L} : H_\alpha^{-1/2}(\partial L) \to H^1_{\alpha,\mathrm{loc}}(\mathbb{R}^3)$ by

$$(SL_{\partial L}\varphi)(x) = \int_{\partial L} G_\alpha(y,x)\,\varphi(y)\,\mathrm{ds}(y)$$

with $x \in \mathbb{R}^3$. The function $SL_{\partial L}\varphi$ is called the *single-layer potential with density* φ on ∂L. The associated *single-layer operator* $S_{\partial L} : H_\alpha^{-1/2}(\partial L) \to H_\alpha^{1/2}(\partial L)$ is defined by $S_{\partial L} = \gamma_D SL_{\partial L}$ where γ_D denotes the trace operator for L.

A proof of the asserted mapping property of $SL_{\partial L}$ can be found e.g. in [53].

Lemma 5.3. *Let R be given as above, i.e. $R = \{x \in \Pi : |x_3| < r\}$ with $r > 0$ such that $\overline{\Omega} \cap \Pi \subset R$ and $\overline{R} \cap \Gamma_i = \overline{R} \cap \Gamma_s = \emptyset$. The single-layer operator $S_{\partial R}$:*

$H_\alpha^{-1/2}(\partial R) \to H_\alpha^{1/2}(\partial R)$ is a Fredholm operator with index zero. Moreover, if k_0^2 is no Dirichlet eigenvalue of $-\Delta$ in R, then $S_{\partial R}$ is injective and has a bounded inverse.

Proof. Clearly, the set R is bounded and Lipschitz. Analog to Theorem 7.6 in [53], it can be shown that $S_{\partial R}$ can be decomposed into $S_{\partial R} = S_0 + (S_{\partial R} - S_0)$ where $S_0 : H_\alpha^{-1/2}(\partial R) \to H_\alpha^{1/2}(\partial R)$ has a coercive real part and $S_{\partial R} - S_0$ is compact. Hence, $S_{\partial R}$ is a Fredholm operator with index zero. To prove the second part of the assertion, assume $\varphi \in \ker S_{\partial R}$. Since $S_{\partial R}\,\varphi = \gamma_D SL_{\partial R}\,\varphi$ and the potential $u_\alpha = SL_{\partial R}\,\varphi$ solves the Helmholtz equation in $\Pi\backslash\partial R$, u_α solves the interior Dirichlet problem (5.3). So, if k_0^2 is not a Dirichlet eigenvalue of $-\Delta$ in R, u_α vanishes in R. In this case, we obtain $\gamma_{D,+}u_\alpha = 0$ on $\partial R \cap \Pi$, where $\gamma_{D,+}$ denotes the trace operator for $\Pi\backslash\overline{R}$. By the uniqueness of a radiating solution to the exterior Dirichlet problem in $\Pi\backslash\overline{R}$, we find that u_α also vanishes in $\Pi\backslash\overline{R}$, hence $u_\alpha \equiv 0$ in Π. The α-quasi-periodicity of u_α and the jump relation $\varphi = -[\gamma_N u_\alpha]_{\partial R}$ then yield $\varphi \equiv 0$. Hence, $S_{\partial R}$ is injective. For a Fredholm operator with index zero, the Fredholm alternative applies, cf. [53]. Therefore, the equation $S_{\partial R}\varphi = g_\alpha$ is uniquely solvable for every $g_\alpha \in H_\alpha^{1/2}(\partial R)$, i.e. $S_{\partial R}$ is bijective, and the inverse $S_{\partial R}^{-1} : H_\alpha^{1/2}(\partial R) \to H_\alpha^{-1/2}(\partial R)$ is bounded (see Corollary 2.12 in [60]). The proof is finished. □

Theorem 5.4. *Assume that k_0^2 is no Dirichlet eigenvalue of $-\Delta$ in R. Let, for all $\delta > 0$, $g_\delta : [0, \|\widetilde{V}_{\Gamma_i}\|^2] \to \mathbb{R}$ be piecewise continuous and such that*

$$|\lambda\, g_\delta(\lambda)| \le C \qquad \text{and} \qquad \lim_{\delta\to 0} g_\delta(\lambda) = \frac{1}{\lambda}$$

for all $\lambda \in (0, \|\widetilde{V}_{\Gamma_i}\|^2]$ and some constant $C > 0$. We define the operators $\widetilde{P}_\delta : H_\alpha^{1/2}(\partial R) \to L^2(\Gamma_i)$ by $\widetilde{P}_\delta = g_\delta(\widetilde{V}_{\Gamma_i}^*\widetilde{V}_{\Gamma_i})\widetilde{V}_{\Gamma_i}^*$, with $g_\delta(\widetilde{V}_{\Gamma_i}^*\widetilde{V}_{\Gamma_i})$ as a functional calculus, as well as $P_\delta : L^2(\Gamma_s) \to L^2(\Gamma_i)$ by $P_\delta = \widetilde{P}_\delta V_{\Gamma_s}$, for $\delta > 0$. Then, $\widetilde{M}P_\delta$ converges to M in the operator norm, i.e. $\|\widetilde{M}P_\delta - M\| \to 0$, for $\delta \to 0$.

Proof. The proof bases on and slightly enhances the proof of Proposition III.12 in [49]. From the strong convergence $\widetilde{M}P_\delta\varphi \to M\varphi$ stated in (5.8) there follows the weak convergence $P_\delta^*\widetilde{M}^*\varphi \rightharpoonup M^*\varphi$ for $\delta \to 0$, for all $\varphi \in L^2(\Gamma_s)$. According to Theorem 4.1 in [25], $\widetilde{P}_\delta$ converges pointwise for $\delta \to 0$ to the (unbounded) inverse of $\widetilde{V}_{\Gamma_i}$ on $\mathcal{R}(\widetilde{V}_{\Gamma_i})$. We recall that $\widetilde{V}_{\Gamma_i}$ is injective and has a dense range according to Lemma 5.1 (ii). Hence, $\{\widetilde{P}_\delta\}_{\delta>0}$ forms a regularization for $\widetilde{V}_{\Gamma_i}$, see

Proposition 3.4 in [25]. The adjoint $\widetilde{P}_\delta^*$ of $\widetilde{P}_\delta$ is represented by

$$\widetilde{P}_\delta^* = \widetilde{V}_{\Gamma_i} g_\delta(\widetilde{V}_{\Gamma_i}^* \widetilde{V}_{\Gamma_i}) = g_\delta(\widetilde{V}_{\Gamma_i} \widetilde{V}_{\Gamma_i}^*) \widetilde{V}_{\Gamma_i} \tag{5.9}$$

(see p. 44 in [25] for an argument) and converges pointwise for $\delta \to 0$ to the (unbounded) inverse of $\widetilde{V}_{\Gamma_i}^*$ on $\mathcal{R}(\widetilde{V}_{\Gamma_i}^*)$. We note that with $\widetilde{V}_{\Gamma_i}$ also $\widetilde{V}_{\Gamma_i}^*$ is injective and has a dense range. With (5.9) and $P_\delta = \widetilde{P}_\delta V_{\Gamma_s}$, we find the identity

$$P_\delta^* \widetilde{M}^* = V_{\Gamma_s}^* g_\delta(\widetilde{V}_{\Gamma_i} \widetilde{V}_{\Gamma_i}^*) \widetilde{V}_{\Gamma_i} \widetilde{M}^*. \tag{5.10}$$

The prior action of $\widetilde{M}^*$ in (5.10) allows to make use of the pointwise convergence of $\widetilde{P}_\delta^*$ for $\delta \to 0$. In fact, the range inclusion $\mathcal{R}(\widetilde{M}^*) \subseteq \mathcal{R}(\widetilde{V}_{\Gamma_i}^*)$ holds. To prove this, we first remark that the preliminary factorization (5.2) implies $\mathcal{R}(\widetilde{M}^*) \subseteq \mathcal{R}((\widetilde{H}_{\Gamma_i}')^*)$. Second, as we are going to show now, the operator $\widetilde{H}_{\Gamma_i}' : L^2(\Gamma_i) \to H_\alpha^1(\Omega)$ can be factorized in the form

$$\widetilde{H}_{\Gamma_i}' = H_{\partial R} S_{\partial R}^{-1} \widetilde{V}_{\Gamma_i}. \tag{5.11}$$

This result leads to $\mathcal{R}((\widetilde{H}_{\Gamma_i}')^*) \subseteq \mathcal{R}(\widetilde{V}_{\Gamma_i}^*)$ and, together with the previous inclusion, to the asserted inclusion. In (5.11), $S_{\partial R} : H_\alpha^{-1/2}(\partial R) \to H_\alpha^{1/2}(\partial R)$ denotes again the single-layer operator on ∂R, which has a bounded inverse according to Lemma 5.3. Now, we define the operator $H_{\partial R} : H_\alpha^{-1/2}(\partial R) \to H_\alpha^1(\Omega)$ by $H_{\partial R}\varphi = (SL_{\partial R}\varphi)|_\Omega$, where R satisfies $\overline{\Omega} \cap \Pi \subset R$ and $\overline{R} \cap \Gamma_i = \overline{R} \cap \Gamma_s = \emptyset$. To establish (5.11), let $\phi \in L^2(\Gamma_i)$ and $\psi = \widetilde{V}_{\Gamma_i}\phi$. For any cuboid $C \subset \Pi$ and $\varphi \in H_\alpha^{-1/2}(B)$ with $B \subset \partial C$, we define the extension $\varphi_{\partial C} \in H_\alpha^{-1/2}(\partial C)$ by

$$\varphi_{\partial C} = \begin{cases} \varphi & \text{on } B \\ 0 & \text{on } \partial C \backslash B \end{cases}.$$

Moreover, by C_i we denote the cuboid which is enclosed by $\partial\Pi$ and the planes which contain $\Gamma_{i,+}$ and $\Gamma_{i,-}$, respectively. Recall that $\Gamma_{i,+}$ and $\Gamma_{i,-}$ are flat and make up the incidence surface $\Gamma_i = \Gamma_{i,+} \cup \Gamma_{i,-}$. Then, we find the identities

$$\gamma_D(SL_{\partial R} S_{\partial R}^{-1}\psi) = \psi = \widetilde{V}_{\Gamma_i}\phi \qquad \text{and} \qquad \gamma_D(SL_{\partial C_i}\phi_{\partial C_i}) = \widetilde{V}_{\Gamma_i}\phi,$$

where γ_D denotes the trace operator for R. From the assumption that k_0^2 is no Dirichlet eigenvalue of $-\Delta$ in R, we conclude that the potentials $SL_{\partial R} S_{\partial R}^{-1}\psi$ and $SL_{\partial C_i}\phi_{\partial C_i}$ are equal in R and so, in particular, in Ω. The fact that $\widetilde{H}_{\Gamma_i}'$ maps ϕ to

$(SL_{\partial C_i}\phi_{\partial C_i})|_\Omega$ then shows (5.11). Returning to (5.10), we note that the physical near field operator $\widetilde{M}$ and its adjoint are compact, and $V^*_{\Gamma_s}$ preserves the pointwise convergence since it is bounded (even compact, too). We are now prepared to apply Theorem 10.6 from [46] and obtain that $P^*_\delta \widetilde{M}^*$ converges in norm to $V^*_{\Gamma_s}(\widetilde{V}^*_{\Gamma_i})^\dagger \widetilde{M}^*$. Since $P^*_\delta \widetilde{M}^*\varphi \rightharpoonup M^*\varphi$ and the weak and the strong limit coincide, there holds $M^* = V^*_{\Gamma_s}(\widetilde{V}^*_{\Gamma_i})^\dagger \widetilde{M}^*$. This finally proves the claim by $\|\widetilde{M}P_\delta - M\| = \|P^*_\delta \widetilde{M}^* - M^*\| \to 0$ for $\delta \to 0$. □

5.2 The electromagnetic case

Many aspects of the approximation of $\mathcal{M}$ by $\widetilde{\mathcal{M}}$ in the electromagnetic case are similar to those in the acoustic case, therefore we skip some details in this section. We adopt here the concept of Section 5 in [44]. To simplify notation below, we define first the auxiliary operators $\widetilde{J} : L^2(\Gamma_i, \mathbb{C}^3) \to L^2_{\mathrm{loc}}(\Pi, \mathbb{C}^3)$ and $J : L^2(\Gamma_s, \mathbb{C}^3) \to L^2_{\mathrm{loc}}(\Pi, \mathbb{C}^3)$ by

$$\widetilde{J}\phi = \int_{\Gamma_i} \mathfrak{G}_\alpha(y,\cdot)\,\phi(y)\,\mathrm{ds}(y) \qquad \text{and} \qquad J\varphi = \int_{\Gamma_s} \overline{\mathfrak{G}_{-\alpha}(y,\cdot)}\,\varphi(y)\,\mathrm{ds}(y).$$

The incident field (4.24) then reads $\widetilde{u}^i_\alpha = (\widetilde{J}\phi)|_{\Pi\setminus\Gamma_i}$, where $\phi \in L^2(\Gamma_i, \mathbb{C}^3)$ is some fixed moment function. The operator $\mathcal{H}_{\Gamma_s}$ defined in (4.28) can be written as $\mathcal{H}_{\Gamma_s}\varphi = \sqrt{|q|}\,\mathcal{H}'_{\Gamma_s}\varphi$ where $\mathcal{H}'_{\Gamma_s}$ is given by $\mathcal{H}'_{\Gamma_s}\varphi = (\operatorname{curl} J\varphi)|_\Omega$ for $\varphi \in L^2(\Gamma_s, \mathbb{C}^3)$. The latter operator is bounded as a mapping from $L^2(\Gamma_s, \mathbb{C}^3)$ to $H_\alpha(\operatorname{curl}, \Omega)$ and, similarly, $\widetilde{\mathcal{H}}'_{\Gamma_i}(\cdot) = (\operatorname{curl}\widetilde{J}(\cdot))|_\Omega$ is bounded from $L^2(\Gamma_i, \mathbb{C}^3)$ to $H_\alpha(\operatorname{curl}, \Omega)$. The near field operators $\mathcal{M}$ and $\widetilde{\mathcal{M}}$ satisfy

$$\widetilde{\mathcal{M}}\phi = \mathcal{G}\sqrt{|q|}\,\widetilde{\mathcal{H}}'_{\Gamma_i}\phi \qquad \text{and} \qquad \mathcal{M}\varphi = \mathcal{G}\sqrt{|q|}\,\mathcal{H}'_{\Gamma_s}\varphi \tag{5.12}$$

for all $\phi \in L^2(\Gamma_i, \mathbb{C}^3)$ and $\varphi \in L^2(\Gamma_s, \mathbb{C}^3)$, where $\mathcal{G}$ is the solution operator defined on p. 62. Again, by R we denote a set $\{x \in \Pi : |x_3| < r\}$ with $r > 0$ such that $\overline{\Omega} \cap \Pi \subset R$ and $\overline{R} \cap \Gamma_i = \overline{R} \cap \Gamma_s = \emptyset$. Finally, we define the operators $\widetilde{\mathcal{V}}'_{\Gamma_i} : L^2(\Gamma_i, \mathbb{C}^3) \to L^2(\partial R, \mathbb{C}^3)$, $\widetilde{\mathcal{V}}_{\Gamma_i} : L^2_t(\Gamma_i) \to L^2_t(\partial R)$, $\mathcal{V}'_{\Gamma_s} : L^2(\Gamma_s, \mathbb{C}^3) \to L^2(\partial R, \mathbb{C}^3)$, and $\mathcal{V}_{\Gamma_s} : L^2_t(\Gamma_s) \to L^2_t(\partial R)$ by

$$\widetilde{\mathcal{V}}'_{\Gamma_i}(\cdot) = (\operatorname{curl}\widetilde{J}(\cdot))\big|_{\partial R}, \qquad \widetilde{\mathcal{V}}_{\Gamma_i} = P^{\partial R}_T\, \widetilde{\mathcal{V}}'_{\Gamma_i}\big|_{L^2_t(\Gamma_i)},$$
$$\mathcal{V}'_{\Gamma_s}(\cdot) = (\operatorname{curl} J(\cdot))\big|_{\partial R}, \qquad \mathcal{V}_{\Gamma_s} = P^{\partial R}_T\, \mathcal{V}'_{\Gamma_s}\big|_{L^2_t(\Gamma_s)},$$

where $P_T^{\partial R} : L^2(\partial R, \mathbb{C}^3) \to L_t^2(\partial R)$, $P_T^{\partial R} : w \mapsto (\nu \times w) \times \nu$, is the tangential projection on ∂R, with ν denoting the outward unit normal to ∂R. Statements analog to those for $\widetilde{V}_{\Gamma_i}$ and V_{Γ_s} in Lemma 5.1 apply to $\widetilde{\mathcal{V}}_{\Gamma_i}$ and $\mathcal{V}_{\Gamma_s}$:

Lemma 5.5.

(i) The operators $\widetilde{\mathcal{V}}_{\Gamma_i}$ and $\mathcal{V}_{\Gamma_s}$ are compact.

(ii) If k_0^2 is not an eigenvalue of the problem

$$\left.\begin{array}{rl} \operatorname{curl}^2 v_\alpha - k_0^2 v_\alpha = 0 & \textit{in } R \\ \gamma_t(\operatorname{curl} v_\alpha) = 0 & \textit{on } \partial R \cap \Pi \end{array}\right\} \tag{5.13}$$

(with an α-quasi-periodic boundary condition on $\partial R \cap \partial\Pi$), then $\widetilde{\mathcal{V}}_{\Gamma_i}$ and $\mathcal{V}_{\Gamma_s}$ are injective and their ranges are dense in $L_t^2(\partial R)$.

(iii) Assume that k_0^2 is not an eigenvalue of the problem (5.13). Let $\{\widetilde{\mathcal{P}}_\delta\}_{\delta>0}$ be a family of operators $\widetilde{\mathcal{P}}_\delta : L_t^2(\partial R) \to L_t^2(\Gamma_i)$ which forms a regularization for the (unbounded) inverse of $\widetilde{\mathcal{V}}_{\Gamma_i}$ on $\mathcal{R}(\widetilde{\mathcal{V}}_{\Gamma_i})$. Then with $\mathcal{P}_\delta = \widetilde{\mathcal{P}}_\delta \mathcal{V}_{\Gamma_s} : L_t^2(\Gamma_s) \to L_t^2(\Gamma_i)$ for $\delta > 0$ there holds the pointwise convergence

$$\widetilde{\mathcal{V}}_{\Gamma_i} \mathcal{P}_\delta \varphi \to \mathcal{V}_{\Gamma_s} \varphi \quad \textit{in } L_t^2(\partial R) \quad \textit{for } \delta \to 0 \tag{5.14}$$

for every $\varphi \in L_t^2(\Gamma_s)$.

Proof.

(i) Since $\overline{R} \cap \Gamma_s = \emptyset$ and the Green's tensor $\mathfrak{G}_{-\alpha}(y,x) = \mathfrak{G}_\alpha(x,y)$ is smooth for all $y \in \Gamma_s$ and all x in an open proper superset of R which does not intersect Γ_s, the mapping $\varphi \mapsto (\operatorname{curl} J\varphi)|_R$ is bounded from $L_t^2(\Gamma_s)$ to $H_\alpha^2(R, \mathbb{C}^3)$. The embedding $H_\alpha^2(R, \mathbb{C}^3) \hookrightarrow H_\alpha^1(R, \mathbb{C}^3)$ is compact, and the tangential components trace operator γ_T is bounded as a mapping from $H_\alpha^1(R, \mathbb{C}^3)$ to $L_t^2(\partial R)$. Hence, the compactness of $\mathcal{V}_{\Gamma_s} : L_t^2(\Gamma_s) \to L_t^2(\partial R)$ follows by the representation $\mathcal{V}_{\Gamma_s}\varphi = \gamma_T((\operatorname{curl} J\varphi)|_R)$. A similar argumentation applies to $\widetilde{\mathcal{V}}_{\Gamma_i}$.

(ii) The proof uses the idea of the proof of Lemma 5.1 in [44]. Again, we confine ourselves to considering the operator $\mathcal{V}_{\Gamma_s}$. Let $g \in \ker \mathcal{V}_{\Gamma_s}$ and define the vector potential $v_\alpha \in H_{\alpha,\mathrm{loc}}(\operatorname{curl}, \Pi)$ by

$$v_\alpha(x) = \operatorname{curl} \int_{\Gamma_s} \overline{\mathfrak{G}_{-\alpha}(y,x)}\, g(y)\, \mathrm{ds}(y) = \operatorname{curl} \int_{\Gamma_s} \overline{G_{-\alpha}(y,x)}\, g(y)\, \mathrm{ds}(y)$$

with $x \in \Pi$. According to the definitions, there holds $(\nu \times v_\alpha|_{\partial R}) \times \nu = \mathcal{V}_{\Gamma_s} g = 0$ and, since $\nu \times v_\alpha|_{\partial R}$ and ν are perpendicular, also $\nu \times v_\alpha|_{\partial R} = 0$. Inspecting Theorem 6.11 in [17] and Theorem 6.11 in [53] reveals that v_α satisfies the jump relations

$$[\gamma_t v_\alpha]_{\Gamma_s} = g \qquad \text{and} \qquad [\gamma_t(\operatorname{curl} v_\alpha)]_{\Gamma_+ \cup \Gamma_-} = 0. \tag{5.15}$$

Moreover, v_α solves the homogeneous Maxwell's equation $\operatorname{curl}^2 v_\alpha - k_0^2 v_\alpha = 0$ in $\Pi \backslash \Gamma_s$ and has an expansion of the form (4.44) in $R_\pm$. In particular, it solves the interior eigenvalue problem

$$\left.\begin{aligned} \operatorname{curl}^2 v_\alpha - k_0^2 v_\alpha &= 0 \quad \text{in } R \\ \nu \times v_\alpha &= 0 \quad \text{on } \partial R \cap \Pi \end{aligned}\right\}. \tag{5.16}$$

Now, we define a function $\widetilde{v}_\alpha$ by $\operatorname{curl} v_\alpha + \mathrm{i} k_0 \widetilde{v}_\alpha = 0$ in R. Due to the homogeneous Maxwell's equation, there holds the complementary equation $\operatorname{curl} \widetilde{v}_\alpha - \mathrm{i} k_0 v_\alpha = 0$ in R. By applying another 'curl' to the first equation in (5.16), we find that $\widetilde{v}_\alpha$ fulfills

$$\left.\begin{aligned} \operatorname{curl}^2 \widetilde{v}_\alpha - k_0^2 \widetilde{v}_\alpha &= 0 \quad \text{in } R \\ \nu \times \operatorname{curl} \widetilde{v}_\alpha &= 0 \quad \text{on } \partial R \cap \Pi \end{aligned}\right\}. \tag{5.17}$$

This is just the interior eigenvalue problem (5.13), hence $\widetilde{v}_\alpha$ has to vanish under the assumption on k_0^2. We conclude that also v_α vanishes in R and, by analyticity, so it does in $\{x \in \Pi : m_- < x_3 < m_+\}$. From this and the second jump relation in (5.15) we obtain $\gamma_{t,+}(\operatorname{curl} v_\alpha) = 0$ on $\Gamma_+ \cup \Gamma_-$, where $\gamma_{t,+}$ is the tangential trace operator for $R_+ \cup R_-$. We are now in a similar situation as in the proof of Proposition 4.5 (ii), with $\widetilde{v}_\alpha$ playing the role of $\widetilde{h}_\alpha$ there, and conclude that $\widetilde{v}_\alpha$ and v_α vanish in $R_+ \cup R_-$. Finally, the first jump relation in (5.15) yields $g \equiv 0$, showing that $\mathcal{V}_{\Gamma_s}$ is injective. To prove the denseness of its range in $L_t^2(\partial R)$, we prove that also the adjoint $\mathcal{V}_{\Gamma_s}^* : L_t^2(\partial R) \to L_t^2(\Gamma_s)$ is injective. First, we observe that

$$\begin{aligned} \langle P_T^{\partial R} g, h \rangle_{\partial R} &= \langle P_T^{\partial R} g, P_T^{\partial R} h + (\mathrm{id} - P_T^{\partial R}) h \rangle_{\partial R} \\ &= \langle P_T^{\partial R} g, P_T^{\partial R} h \rangle_{\partial R} \\ &= \langle P_T^{\partial R} g + (\mathrm{id} - P_T^{\partial R}) g, P_T^{\partial R} h \rangle_{\partial R} \\ &= \langle g, P_T^{\partial R} h \rangle_{\partial R} \end{aligned}$$

holds for all $g, h \in L^2(\partial R, \mathbb{C}^3)$, where $\langle \cdot, \cdot \rangle_{\partial R}$ denotes the usual scalar product for this space. Of course, a similar relation applies to the tangential projection $P_T^{\Gamma_s} : L^2(\Gamma_s, \mathbb{C}^3) \to L_t^2(\Gamma_s)$ with respect to the scalar product $\langle \cdot, \cdot \rangle_{\Gamma_s}$ in

$L^2(\Gamma_s,\mathbb{C}^3)$. Using this, we derive that

$$\begin{aligned}\langle \mathcal{V}_{\Gamma_s} g, h\rangle_{\partial R} &= \langle \mathcal{V}_{\Gamma_s} P_T^{\Gamma_s} g, h\rangle_{\partial R} \\ &= \langle P_T^{\partial R} \mathcal{V}'_{\Gamma_s} P_T^{\Gamma_s} g, h\rangle_{\partial R} \\ &= \langle g, P_T^{\Gamma_s} (\mathcal{V}'_{\Gamma_s})^* P_T^{\partial R} h\rangle_{\Gamma_s} \\ &= \langle g, P_T^{\Gamma_s} (\mathcal{V}'_{\Gamma_s})^* h\rangle_{\Gamma_s}\end{aligned}$$

holds for all $g \in L_t^2(\Gamma_s)$ and $h \in L_t^2(\partial R)$. Hence, the adjoint of $\mathcal{V}_{\Gamma_s}$ is given by $\mathcal{V}_{\Gamma_s}^* = P_T^{\Gamma_s}(\mathcal{V}'_{\Gamma_s})^*|_{L_t^2(\partial R)}$, and $(\mathcal{V}'_{\Gamma_s})^* : L^2(\partial R,\mathbb{C}^3) \to L^2(\Gamma_s,\mathbb{C}^3)$ is easily found to be

$$((\mathcal{V}'_{\Gamma_s})^*\varphi)(x) = \operatorname{curl}\int_{\partial R} \mathfrak{G}_{-\alpha}(x,y)\,\varphi(y)\,\mathrm{ds}(y)$$

with $\varphi \in L^2(\partial R,\mathbb{C}^3)$ and $x \in \Gamma_s$. Now, we choose $h \in \ker \mathcal{V}_{\Gamma_s}^*$ and define $w_\alpha \in H_{\alpha,\mathrm{loc}}(\operatorname{curl}, 2\cdot\Pi)$ by

$$w_\alpha(x) = \operatorname{curl}\int_{\partial R} \mathfrak{G}_{-\alpha}(x,y)\,h(y)\,\mathrm{ds}(y) = \operatorname{curl}\int_{\partial R} G_{-\alpha}(x,y)\,h(y)\,\mathrm{ds}(y)$$

with $x \in 2\cdot\Pi$, such that $(\nu \times w_\alpha|_{\Gamma_s}) \times \nu = \mathcal{V}_{\Gamma_s}^* h = 0$ and also $\nu \times w_\alpha|_{\Gamma_s} = 0$. Similar to (5.15), there hold the jump relations

$$[\gamma_t w_\alpha]_{\partial R} = h \qquad \text{and} \qquad [\gamma_t(\operatorname{curl} w_\alpha)]_{\partial R} = 0. \tag{5.18}$$

By the analyticity of $w_\alpha((\cdot,\cdot,x_3))$ in $\Gamma^\star = \{x \in 2\cdot\Pi : x_3 = m_- \vee x_3 = m_+\}$ and $\nu = \pm e_3$ on $\Gamma_{s,\pm}$, from $\nu \times w_\alpha|_{\Gamma_s} = 0$ we gain $\nu \times w_\alpha|_{\Gamma^\star} = 0$. In particular, w_α solves the exterior problem

$$\left.\begin{aligned}\operatorname{curl}^2 w_\alpha - k_0^2 w_\alpha &= 0 \quad \text{in } R^\star \\ \nu \times w_\alpha &= 0 \quad \text{on } \Gamma^\star\end{aligned}\right\} \tag{5.19}$$

where $R^\star = \{x \in 2\cdot\Pi : x_3 < m_- \vee x_3 > m_+\}$, together with an α-quasi-periodic boundary condition (still with respect to Π, i.e. to the quasi-period $\Lambda = (2\pi, 2\pi, 0)^T$) on $\overline{R^\star} \cap \partial(2\cdot\Pi)$. By (5.19) and the smoothness of w_α in $\{x \in 2\cdot\Pi : |x_3| > r\}$, we obtain

$$\left.\begin{aligned}\Delta w_\alpha + k_0^2 w_\alpha &= 0 \quad && \text{in } R^\star \\ \operatorname{div} w_\alpha &= 0 && \text{on } \Gamma^\star \\ \nu \times w_\alpha &= 0 && \text{on } \Gamma^\star\end{aligned}\right\}. \tag{5.20}$$

Accompanied by radiating behavior of w_α, this is a special, quasi-periodic *exterior electric boundary value problem*, cp. the proof of Proposition 4.5 (ii). Using the Rayleigh expansion of the component functions of w_α and the boundary conditions on $\Gamma^\star$, one easily proves that (5.20) has at most one solution. Hence, w_α vanishes in $R^\star$ and, by analyticity, so it does in all of $\{x \in 2\cdot\Pi : |x_3| > r\}$. The second jump relation in (5.18) then implies $\gamma_{t,-}(\operatorname{curl} w_\alpha) = 0$ on $\partial R \cap \Pi$, where $\gamma_{t,-}$ is the tangential trace operator for R. Moreover, w_α satisfies an α-quasi-periodic boundary condition on $\partial R \cap \partial\Pi$ and solves $\operatorname{curl}^2 w_\alpha - k_0^2 w_\alpha = 0$ in R. Thus, it is a solution to the interior eigenvalue problem (5.13) and vanishes in R under the assumption on k_0^2. The α-quasi-periodicity of w_α and the first jump relation in (5.18) now reveal $h \equiv 0$. The proof is complete.

(iii) This is a consequence of standard regularization theory, see, e.g., [25]. □

In the following, we assume that the condition in Lemma 5.5 (ii) is fulfilled, i.e. k_0^2 is no eigenvalue of the problem

$$\left.\begin{aligned} \operatorname{curl}^2 v_\alpha - k_0^2 v_\alpha &= 0 && \text{in } R \\ \gamma_t(\operatorname{curl} v_\alpha) &= 0 && \text{on } \partial R \cap \Pi \end{aligned}\right\}. \tag{5.21}$$

The inhomogeneous eigenvalue problem, which corresponds to the problem (5.21) with $\gamma_t(\operatorname{curl} v_\alpha) = f_\alpha$ on $\partial R \cap \Pi$ with arbitrary $f_\alpha \in Y_\alpha(\partial R)$ instead, can be formulated in the variational sense via the Green's identity (2.25). The volume integral part of this formulation defines a sesquilinear form on $H_\alpha(\operatorname{curl}, R)$. It can be shown that this form induces a Fredholm operator with index zero, cf. [14, Section 3]. Due to the above condition on k_0^2, this operator is injective. Moreover, the boundary integral part constitutes a bounded conjugate-linear functional on $H_\alpha(\operatorname{curl}, R)$. In consequence, there is a unique solution to the inhomogeneous eigenvalue problem and the solution depends continuously on the tangential trace data. Besides, we remark that there is a slight connection between the eigenvalue problem (5.21) and the problem

$$\left.\begin{aligned} \Delta w_\alpha + k_0^2 w_\alpha &= 0 && \text{in } R \\ \gamma_D(\operatorname{div} w_\alpha) &= 0 && \text{on } \partial R \cap \Pi \\ \gamma_t w_\alpha &= 0 && \text{on } \partial R \cap \Pi \end{aligned}\right\}, \tag{5.22}$$

which is an α-quasi-periodic variant of the so-called *interior electric boundary value problem*, cf. [16]. The latter one can be derived from the former one by applying another 'curl' to the homogeneous Maxwell's equation in (5.21) and defining w_α by the equation $\operatorname{curl} v_\alpha + \mathrm{i}k_0 w_\alpha = 0$ in R. Obviously, if $\operatorname{curl}^2 v_\alpha - k_0^2 v_\alpha$ holds in R, then there also applies the complementary equation $\operatorname{curl} w_\alpha - \mathrm{i}k_0 v_\alpha = 0$ in R. From this we conclude that if k_0^2 is an eigenvalue of the problem (5.21), then it is an eigenvalue of the problem (5.22).

Now, for $\varphi \in L^2(\Gamma_s, \mathbb{C}^3)$, both α-quasi-periodic vector potentials

$$x \longmapsto (\widetilde{J}\mathcal{P}_\delta\varphi)(x) = \int_{\Gamma_i} \mathfrak{G}_\alpha(y,x)\,(\mathcal{P}_\delta\varphi)(y)\,\mathrm{ds}(y) \qquad \text{and}$$

$$x \longmapsto (J\varphi)(x) = \int_{\Gamma_s} \overline{\mathfrak{G}_{-\alpha}(y,x)}\,\varphi(y)\,\mathrm{ds}(y)$$

satisfy the homogeneous Maxwell's equation in R. Since the pointwise convergence (5.14) implies

$$\nu \times \widetilde{\mathcal{V}}'_{\Gamma_i}\mathcal{P}_\delta\,\varphi \to \nu \times \mathcal{V}'_{\Gamma_s}\varphi \quad \text{in } L^2_t(\partial R) \quad \text{for } \delta \to 0,$$

the continuous dependence of the unique solution to the problem (5.21) from the tangential trace data yields that $\widetilde{\mathcal{H}}'_{\Gamma_i}\mathcal{P}_\delta\,\varphi$ converges to $\mathcal{H}'_{\Gamma_s}\varphi$ in $H_\alpha(\operatorname{curl},\Omega)$ as δ goes to zero. Finally, we arrive at the pointwise convergence

$$\widetilde{\mathcal{M}}\mathcal{P}_\delta\,\varphi = \mathcal{G}\sqrt{|q|}\,\widetilde{\mathcal{H}}'_{\Gamma_i}\mathcal{P}_\delta\,\varphi \to \mathcal{G}\sqrt{|q|}\,\mathcal{H}'_{\Gamma_s}\varphi = \mathcal{M}\varphi \qquad \text{for } \delta \to 0 \tag{5.23}$$

for every $\varphi \in L^2(\Gamma_s, \mathbb{C}^3)$. In principle, we can compute $\mathcal{M}\varphi$ to a prescribed accuracy for any given density $\varphi \in L^2(\Gamma_s, \mathbb{C}^3)$. As in the acoustic case, it will become crucial later that $\widetilde{\mathcal{M}}\mathcal{P}_\delta$ converges to $\mathcal{M}$ even in the operator norm. We prove now that this in fact holds.

Theorem 5.6. *Assume that k_0^2 is no eigenvalue of the problem (5.21). Let, for all $\delta > 0$, $g_\delta : [0, \|\widetilde{\mathcal{V}}'_{\Gamma_i}\|^2] \to \mathbb{R}$ be piecewise continuous and fulfill*

$$|\lambda\, g_\delta(\lambda)| \le C \qquad \textit{and} \qquad \lim_{\delta \to 0} g_\delta(\lambda) = \frac{1}{\lambda}$$

for all $\lambda \in (0, \|\widetilde{\mathcal{V}}'_{\Gamma_i}\|^2]$ and some constant $C > 0$. Then the operator family $\{\widetilde{\mathcal{P}}_\delta\}_{\delta>0}$ with $\widetilde{\mathcal{P}}_\delta = g_\delta((\widetilde{\mathcal{V}}'_{\Gamma_i})^*\widetilde{\mathcal{V}}'_{\Gamma_i})(\widetilde{\mathcal{V}}'_{\Gamma_i})^* : L^2(\partial R, \mathbb{C}^3) \to L^2(\Gamma_i, \mathbb{C}^3)$ forms a regularization for the (unbounded) inverse of $\widetilde{\mathcal{V}}'_{\Gamma_i}$ on $\mathcal{R}(\widetilde{\mathcal{V}}'_{\Gamma_i})$. With $\mathcal{P}_\delta = \widetilde{\mathcal{P}}_\delta \mathcal{V}'_{\Gamma_s}$ for $\delta > 0$, $\widetilde{\mathcal{M}}\mathcal{P}_\delta$ converges to $\mathcal{M}$ in the operator norm as δ goes to zero.

Proof. It is sufficient to show that the range inclusion $\mathcal{R}((\widetilde{\mathcal{H}}'_{\Gamma_i})^*) \subseteq \mathcal{R}((\widetilde{\mathcal{V}}'_{\Gamma_i})^*)$ holds. Then an argumentation similar to that in the proof of Theorem 5.4 verifies the assertion. To start, we recall that $\widetilde{\mathcal{H}}'_{\Gamma_i} : L^2(\Gamma_i, \mathbb{C}^3) \to L^2(\Omega, \mathbb{C}^3)$ is given by

$$\widetilde{\mathcal{H}}'_{\Gamma_i}\phi = (\operatorname{curl}\widetilde{J}\phi)\big|_\Omega = \operatorname{curl}\int_{\Gamma_i} G_\alpha(y,\cdot)\,\phi(y)\,\mathrm{ds}(y). \tag{5.24}$$

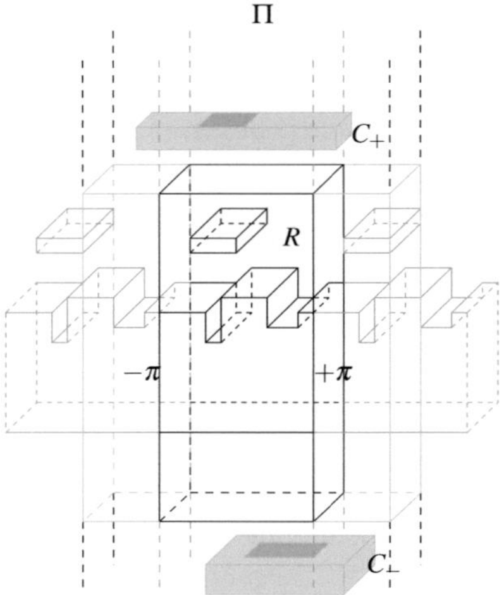

Figure 5.2: Example for $C_\pm$

Letting aside the 'curl' in this expression, the remainder has the form of the operator $\widetilde{H}'_{\Gamma_i}$ from the acoustic case, discussed in Section 5.1, but is applied here to a vectorial argument $\phi \in L^2(\Gamma_i, \mathbb{C}^3)$. We denote this operator by $\mathcal{H}^a_{\Gamma_i}$, it is bounded from $L^2(\Gamma_i, \mathbb{C}^3)$ to $H^1_\alpha(\Omega, \mathbb{C}^3)$ and, because of $\Omega \cap \Gamma_i = \emptyset$, also to $H^2_\alpha(\Omega, \mathbb{C}^3)$. Since the action of $\mathcal{H}^a_{\Gamma_i}$ on ϕ does not involve any mixing of the components of ϕ, $\mathcal{H}^a_{\Gamma_i}$ allows a factorization in the form

$$\mathcal{H}^a_{\Gamma_i} = \mathcal{H}^a_{\partial R} (S^a_{\partial R})^{-1} \widetilde{\mathcal{V}}^a_{\Gamma_i}, \tag{5.25}$$

where the operators on the right-hand side are the vectorial counterparts of the corresponding operators in the factorization (5.11). Since the adjoint of $\widetilde{\mathcal{H}}'_{\Gamma_i}$ is given by $(\widetilde{\mathcal{H}}'_{\Gamma_i})^* g = \operatorname{curl} (\mathcal{H}^a_{\Gamma_i})^* g$ for $g \in L^2(\Omega, \mathbb{C}^3)$, by means of (5.25) we obtain

$$(\widetilde{\mathcal{H}}'_{\Gamma_i})^* g = \operatorname{curl} (\widetilde{\mathcal{V}}^a_{\Gamma_i})^* \mathcal{X}^* g \tag{5.26}$$

where $\mathcal{X} = \mathcal{H}^a_{\partial R}(S^a_{\partial R})^{-1} : H^{1/2}_\alpha(\partial R, \mathbb{C}^3) \to H^1_\alpha(\Omega, \mathbb{C}^3)$ is bounded. The operator $(\widetilde{\mathcal{V}}^a_{\Gamma_i})^*$ is given by $(\widetilde{\mathcal{V}}^a_{\Gamma_i})^* \phi = (\widetilde{SL}_{\partial R} \phi)|_{\Gamma_i}$ for $\phi \in H_\alpha^{-1/2}(\partial R, \mathbb{C}^3)$ where $\widetilde{SL}_{\partial R} \phi$ denotes the single-layer vector potential

$$(\widetilde{SL}_{\partial R} \phi)(x) = \int_{\partial R} \overline{G_\alpha(x,y)}\, \phi(y)\, \mathrm{ds}(y), \qquad x \in \mathbb{R}^3. \tag{5.27}$$

This potential solves the vector Helmholtz equation in $(2\cdot\Pi)\setminus\partial R$, see the first comment on p. 138 in [62], and thus is smooth there. Now, think of $\Gamma_{i,\pm}$ as part of a closed Lipschitz surface $C_\pm$ in $\{x\in 2\cdot\Pi : |x_3|>r\}$, respectively, as illustrated in Figure 5.2. The 'curl' of the potential (5.27) lies in $H^1_{\alpha,\mathrm{loc}}((2\cdot\Pi)\setminus\overline{R})$ and hence has a well-defined Dirichlet trace on $C_\pm$ in $H^{1/2}_\alpha(C_\pm,\mathbb{C}^3)$, respectively. Now, restricting the trace again to Γ_i, it follows that $g\mapsto \operatorname{curl}(\widetilde{\mathcal{V}}^a_{\Gamma_i})^*\mathcal{X}^*g$ is well-defined as a mapping from $L^2(\Omega,\mathbb{C}^3)$ to $L^2(\Gamma_i,\mathbb{C}^3)$. Finally, there holds

$$\begin{aligned}(\widetilde{\mathcal{H}}'_{\Gamma_i})^*g &= \operatorname{curl}(\widetilde{\mathcal{V}}^a_{\Gamma_i})^*\mathcal{X}^*g\\ &= \operatorname{curl}\int_{\partial R}\overline{G_\alpha(\cdot,y)}\,(\mathcal{X}^*g)(y)\,\mathrm{d}s(y)\\ &= \operatorname{curl}\int_{\partial R}\overline{\mathfrak{G}_\alpha(\cdot,y)}\,(\mathcal{X}^*g)(y)\,\mathrm{d}s(y),\end{aligned}$$

which reveals $\mathcal{R}((\widetilde{\mathcal{H}}'_{\Gamma_i})^*)\subseteq\mathcal{R}((\widetilde{\mathcal{V}}'_{\Gamma_i})^*)$. □

Chapter 6

Reconstruction of the shape

In Chapters 3 and 4, we set up and discussed the factorization of the acoustic and electromagnetic artificial near field operators M and $\mathcal{M}$. Now, we consider the identification of the shape of the periodic medium based on these factorizations, which we recall to be

$$M = H_{\Gamma_s}^* T H_{\Gamma_s} \qquad \text{and} \qquad \mathcal{M} = \mathcal{H}_{\Gamma_s}^* \mathcal{T} \mathcal{H}_{\Gamma_s}, \tag{6.1}$$

see (3.39) and (4.42). We first restate the theorems which form the abstract foundation of the Factorization Method and have been proposed by KIRSCH in the papers [39, 42], with one important enhancement by LECHLEITER, cf. Theorem I.7 in [49]. In Subsections 3.2.3 and 4.2.3 it was shown how the type of the contrast determines properties of the inner operators T and $\mathcal{T}$. These in turn play a central role in the functional analytic result behind the Factorization Method, which we formulate in Section 6.2. It can be applied for a decent class of contrasts in our inverse scattering problem. For almost everywhere absorbing media one can use a corollary to this result. The reconstruction of such media and those described by the more general contrasts is then made concrete in Sections 6.3 and 6.4.

6.1 Range tests

The next two results illustrate an immanent link between the shape of the scattering medium and the range of the first operator (which cannot be evaluated without the knowledge of Ω) in the factorization of M and $\mathcal{M}$, respectively. The statements and the proofs of this link are very similar in the acoustic and the electromagnetic case.

Theorem 6.1. *For any $z \in \Pi$ we define $\psi_{\alpha,z} \in L^2(\Gamma_s)$ by*

$$\psi_{\alpha,z}(x) = G_\alpha(z,x), \qquad x \in \Gamma_s.$$

There holds $z \in \Omega$ if and only if $\psi_{\alpha,z} \in \mathcal{R}(H_{\Gamma_s}^*)$.

Proof. The proof follows the idea of the proof of Theorem 4.6 in [45]. Assume first that $z \in \Omega$. Let $\varepsilon > 0$ be such that the closed ball $\overline{B(z,\varepsilon)}$ with center z and radius ε is contained in Ω. We choose a mollifier $\phi \in C^\infty(\mathbb{R})$ with $\phi(t) = 1$ for $|t| \geq \varepsilon$ and $\phi(t) = 0$ for $|t| \leq \varepsilon/2$, and define $v_\alpha \in C^\infty_\alpha(\Pi)$ by $v_\alpha(x) = \phi(|x - z|)\, G_\alpha(z,x)$. In $\Pi \backslash \overline{B(z,\varepsilon)}$, it satisfies $v_\alpha = G_\alpha(z,\cdot)$ and thus $\Delta v_\alpha + k_0^2 v_\alpha = 0$. Since $G_\alpha(z,\cdot)$ obeys the radiation condition (3.18), v_α radiates according to (3.5). Therefore, by Green's formula (3.20) with $L = \Omega$, we obtain

$$\begin{aligned} v_\alpha(x) &= \int_{\partial\Omega\cap\Pi} \left(G_\alpha(y,x) \frac{\partial}{\partial \nu_y} G_\alpha(z,y) - G_\alpha(z,y) \frac{\partial}{\partial \nu_y} G_\alpha(y,x) \right) \mathrm{d}s(y) - \\ &\quad - \int_{|y-z|<\varepsilon} (\Delta v_\alpha(y) + k_0^2 v_\alpha(y))\, G_\alpha(y,x)\, \mathrm{d}y \\ &= - \int_{|y-z|<\varepsilon} (\Delta v_\alpha(y) + k_0^2 v_\alpha(y))\, G_\alpha(y,x)\, \mathrm{d}y \end{aligned}$$

for $x \in \Omega$, where the first integral vanishes by formula (3.21). Using an analytic continuation argument for v_α in Ω^{ext}, we conclude that

$$\psi_{\alpha,z} = v_\alpha = - \int_{|y-z|<\varepsilon} (\Delta v_\alpha(y) + k_0^2 v_\alpha(y))\, G_\alpha(y,\cdot)\, \mathrm{d}y \qquad \text{on } \Gamma_s. \tag{6.2}$$

Now, we define

$$g = \begin{cases} -(\Delta v_\alpha + k_0^2 v_\alpha)/\sqrt{|q|} & \text{in } \overline{B(z,\varepsilon)} \\ 0 & \text{in } \Omega \backslash \overline{B(z,\varepsilon)} \end{cases}. \tag{6.3}$$

It is $g \in L^2(\Omega)$ since $|q|$ is locally bounded from below in $\Omega \supset \overline{B(z,\varepsilon)}$, cf. Assumptions 3.2. Then, by (6.2) and the form of the operator $H^*_{\Gamma_s}$ from (3.30), we find $\psi_{\alpha,z} = H^*_{\Gamma_s} g$, i.e. $\psi_{\alpha,z} \in \mathcal{R}(H^*_{\Gamma_s})$.

Let now $z \in \Pi \backslash \Omega$ and assume, on the contrary, that there exists some $g \in L^2(\Omega)$ such that $H^*_{\Gamma_s} g = \psi_{\alpha,z} = G_\alpha(z,\cdot)$ on Γ_s. Then, by the one-to-one correspondence in $\Omega^{\mathrm{ext}} = \Pi \backslash \overline{\Omega}$ between radiating solutions to the Helmholtz equation and their near fields on Γ_s, we obtain

$$\int_\Omega G_\alpha(y,x)\, g(y)\, \sqrt{|q(y)|}\, \mathrm{d}y = G_\alpha(z,x), \qquad x \in \Pi \backslash (\Omega \cup \{z\}). \tag{6.4}$$

Here, we also used that Ω' has no inclusions of the background medium and the function defined by the left-hand side has a continuous extension from Ω^{ext} to $\partial\Omega \cap \Pi$. This function even lies in $C^1_\alpha(\Pi)$, cf. Lemma 4.1 in [27] together with the decomposition (3.17) of G_α. Moreover, it is a solution to the Helmholtz equation, and thus analytic, in Ω^{ext}. The right-hand side, however, has a singularity at $z \notin \Omega$, which leads to a contradiction. We conclude that $\psi_{\alpha,z} \notin \mathcal{R}(H^*_{\Gamma_s})$. □

The counterpart of Theorem 6.1 for the electromagnetic case has been proven for a related setting in [44, Theorem 4.2]. The proof for our setting could be done in an analogous manner with the aid of Proposition 4.3. However, we give an alternative proof here, which uses our representation Theorem 4.1 and reveals the same structure as the above proof for the acoustic case.

Theorem 6.2. *For any $z \in \Pi$ and fixed $p \in \mathbb{R}^3$ we define $\widetilde{\psi}_{\alpha,z} \in L^2(\Gamma_s, \mathbb{C}^3)$ by*

$$\widetilde{\psi}_{\alpha,z}(x) = k_0^2\, \mathfrak{G}_\alpha(z,x)\, p, \qquad x \in \Gamma_s.$$

There holds $z \in \Omega$ if and only if $\widetilde{\psi}_{\alpha,z} \in \mathcal{R}(\mathcal{H}^*_{\Gamma_s})$.

Proof. We assume first that $z \in \Omega$ and let $\varepsilon > 0$ be such that the closed ball $\overline{B(z,\varepsilon)}$ is contained in Ω. In addition, let $\phi \in C^\infty(\mathbb{R})$ be such that $\phi(t) = 1$ for $|t| \geq \varepsilon$ and $\phi(t) = 0$ for $|t| \leq \varepsilon/2$. Then, we define $v_\alpha \in C^\infty_\alpha(\Pi, \mathbb{C}^3)$ by $v_\alpha(x) = \operatorname{curl}^2(\phi(|x-z|)\, \mathfrak{G}_\alpha(z,x)\, p)$. In $\Pi \backslash \overline{B(z,\varepsilon)}$, it satisfies $v_\alpha = \operatorname{curl}^2(\mathfrak{G}_\alpha(z,\cdot)\, p) = k_0^2\, \mathfrak{G}_\alpha(z,\cdot)\, p$ and, in particular, $v_\alpha|_{\Gamma_s} = \widetilde{\psi}_{\alpha,z}$. Since $\mathfrak{G}_\alpha(z,\cdot)$ obeys the radiation condition (4.16), v_α radiates according to (4.5). By Theorem 4.1 with $L = \Omega$, we obtain

$$\begin{aligned} v_\alpha(x) &= -\int_{\partial\Omega\cap\Pi} \big(\mathfrak{G}_\alpha(\cdot,x)\,(\nu \times \operatorname{curl} v_\alpha) - \operatorname{curl}_y \mathfrak{G}_\alpha(\cdot,x)\,(\nu \times v_\alpha)\big)\,\mathrm{d}s(y) + \\ &\quad + \int_{|y-z|<\varepsilon} \mathfrak{G}_\alpha(\cdot,x)\,(\operatorname{curl}^2 v_\alpha - k_0^2\, v_\alpha)\,\mathrm{d}y \\ &= \int_{|y-z|<\varepsilon} \mathfrak{G}_\alpha(\cdot,x)\,(\operatorname{curl}^2 v_\alpha - k_0^2\, v_\alpha)\,\mathrm{d}y \end{aligned} \tag{6.5}$$

for $x \in \Omega$, where the first integral vanishes by formula (4.19) and $v_\alpha = k_0^2\, \mathfrak{G}_\alpha(z,\cdot)\, p$ on $\partial\Omega \cap \Pi$. By analyticity of v_α in $\Pi \backslash \overline{B(z,\varepsilon)}$, the representation (6.5) holds in fact in all of Π. Now, by a similar argument as in the proof of Theorem 6.1, we find that $\widetilde{\psi}_{\alpha,z}$ is in the range of the operator

$$g \longmapsto \int_\Omega \mathfrak{G}_\alpha(y,x)\, g(y)\, \sqrt{|q(y)|}\,\mathrm{d}y, \quad x \in \Gamma_s.$$

However, comparing the form of the operator $\mathcal{H}^*_{\Gamma_s}$ from (4.29), the proof is not yet finished. We note that in a neighborhood $\widetilde{\Gamma}$ of Γ_s (in Π) we can reformulate

v_α as

$$\begin{aligned} v_\alpha(x) &= \int_{|y-z|<\varepsilon} \frac{1}{k_0^2} \big(\operatorname{curl}_y^2 \mathfrak{G}_\alpha(y,x)\big)\big(\operatorname{curl}^2 v_\alpha(y) - k_0^2 v_\alpha(y)\big)\,\mathrm{d}y \\ &= \int_{|y-z|<\varepsilon} \frac{1}{k_0^2} \operatorname{curl}_x\big(-\operatorname{curl}_y \mathfrak{G}_\alpha(y,x)\big)\big(\operatorname{curl}^2 v_\alpha(y) - k_0^2 v_\alpha(y)\big)\,\mathrm{d}y \\ &= \operatorname{curl} \int_\Omega \big(\operatorname{curl}_y \mathfrak{G}_\alpha(y,x)\big)\, \widetilde{g}_\alpha(y)\,\mathrm{d}y \end{aligned}$$

for $x \in \widetilde{\Gamma}$, where $\widetilde{g}_\alpha$ is defined by

$$\widetilde{g}_\alpha = \begin{cases} -k_0^{-2} \,(\operatorname{curl}^2 v_\alpha - k_0^2 v_\alpha) & \text{in } \overline{B(z,\varepsilon)} \\ 0 & \text{in } \Omega \backslash \overline{B(z,\varepsilon)} \end{cases}.$$

It is easy to see that $\widetilde{g}_\alpha$ is in $C_\alpha^\infty(\Omega, \mathbb{C}^3)$, and thus we can apply the identity (2.20) to deduce that

$$\int_\Omega \big(\operatorname{curl}_y \mathfrak{G}_\alpha(y,x)\big)\, \widetilde{g}_\alpha(y)\,\mathrm{d}y = \int_\Omega \mathfrak{G}_\alpha(y,x) \operatorname{curl} \widetilde{g}_\alpha(y)\,\mathrm{d}y$$

holds for $x \in \widetilde{\Gamma}$ since $\gamma_D \widetilde{g}_\alpha = 0$ on $\partial\Omega$. Hence, for $g \in L^2(\Omega, \mathbb{C}^3)$ defined by

$$g = \operatorname{curl} \widetilde{g}_\alpha / \sqrt{|q|} = \begin{cases} (\operatorname{curl} v_\alpha - k_0^{-2} \operatorname{curl}^3 v_\alpha)/\sqrt{|q|} & \text{in } \overline{B(z,\varepsilon)} \\ 0 & \text{in } \Omega \backslash \overline{B(z,\varepsilon)} \end{cases}$$

we finally obtain $\widetilde{\psi}_{\alpha,z} = \mathcal{H}_{\Gamma_s}^* g$, showing $\widetilde{\psi}_{\alpha,z} \in \mathcal{R}(\mathcal{H}_{\Gamma_s}^*)$.

Let now $z \in \Pi \backslash \Omega$ and assume that there is a $g \in L^2(\Omega, \mathbb{C}^3)$ such that $\widetilde{\psi}_{\alpha,z} = \mathcal{H}_{\Gamma_s}^* g$. By the one-to-one correspondence in Ω^{ext} between radiating solutions to the homogeneous Maxwell's equation $\operatorname{curl}^2 v_\alpha - k_0^2 v_\alpha = 0$ and their near fields on Γ_s, we conclude that

$$\operatorname{curl} \int_\Omega \mathfrak{G}_\alpha(y,x)\, g(y)\, \sqrt{|q(y)|}\,\mathrm{d}y = k_0^2\, \mathfrak{G}_\alpha(z,\cdot)\, p, \qquad x \in \Omega^{\text{ext}} \backslash \{z\}.$$

Again, we used that Ω' has no inclusions of the background medium. This leads to a contradiction since by Proposition 4.3 (ii) there holds $v_\alpha|_B \in H_\alpha(\operatorname{curl}, B)$ for any ball $B \subset \Pi$ which contains z, whereas $(k_0^2 \mathfrak{G}_\alpha(z,\cdot)\, p)|_{B^e} \notin L^2(B^e, \mathbb{C}^3)$ for $B^e = B \cap \Omega^{\text{ext}}$ by the strong singularity at z. This completes the proof. □

In connection with the range identities discussed below, Theorems 6.1 and 6.2 will make it possible to identify the shape of the scattering medium by known data, rather than by $H^*_{\Gamma_s}$ and $\mathcal{H}^*_{\Gamma_s}$, respectively. On the other hand, the regularity arguments used in the proofs above make clear why the Factorization Method in the form we use here is incapable of reconstructing the contrast q as a function. As mentioned already, it is, however, sufficient in many applications to recover the support of q.

6.2 Range identity

We now state the functional analytic result behind the Factorization Method, which for certain factorizations $A = B^* C B$ establishes the identity of the ranges of the operator B^* and of a self-adjoint operator which can be computed (in a quite simple way) from the factorized operator A. As usual, B^* denotes the adjoint of B. The properties of the operators involved in (6.1) which have been proven in the previous chapters will show that the factorizations of the near field operators M and $\mathcal{M}$ match the required pattern. In [49], Theorem I.7, LECHLEITER refined a result from [42] which asserts a range identity of the described type. We will use the following, completely analogous extension of Theorem 2.15 from [45]. We do not reproduce its elaborate proof here.

Theorem 6.3. *Let $X \subseteq U \subseteq X^*$ be a Gelfand triple with Hilbert space U and reflexive Banach space X such that the embedding is dense. Furthermore, let H be a second Hilbert space and let $A : H \to H$, $B : H \to X$, and $C : X \to X^*$ be linear and bounded operators such that*

$$A = B^* C B.$$

We make the following assumptions:

(i) B is compact and injective.

(ii) For some $t \in [0, 2\pi)$ the operator $\mathrm{Re}\,(e^{\mathrm{i}t}C)$ has the form $\mathrm{Re}\,(e^{\mathrm{i}t}C) = \widetilde{C} + K$ with some coercive operator $\widetilde{C}$ and some compact operator K from X to X^*.

(iii) $\mathrm{Im}\,C$ is non-negative on X, i.e. $\langle \mathrm{Im}\,C\phi, \phi \rangle \geq 0$ holds for all $\phi \in X$.

Moreover, we assume that one of the next conditions is fulfilled.

(iv)-a C is injective.

(iv)-b $\mathrm{Im}\,C$ is positive on the (finite-dimensional) null space of $\mathrm{Re}\,(e^{\mathrm{i}t}C)$, that is $\langle \mathrm{Im}\,C\phi, \phi \rangle > 0$ holds for all $\phi \neq 0$ with $\mathrm{Re}\,(e^{\mathrm{i}t}C)\phi = 0$.

Then the operator $A_\sharp = |\mathrm{Re}\,(e^{\mathrm{i}t}A)| + \mathrm{Im}A$ *is positive, and the ranges of* $B^* : X \to H$ *and* $A_\sharp^{1/2} : H \to H$ *coincide.*

Here, $A_\sharp^{1/2}$ denotes the square root of the operator $A_\sharp$. Every non-negative operator $G \in \mathcal{L}(H)$ on a complex Hilbert space H has a unique non-negative square root $G^{1/2} \in \mathcal{L}(H)$, which solves $S^2 = G$, cf. [60, Theorem 12.33]. The absolute value $|G|$ of any operator $G \in \mathcal{L}(H)$ is defined as $|G| = (G^*G)^{1/2}$, see, e.g., p. 329 in [70]. Since, in Theorem 6.3, B is compact and C is bounded, A, $\mathrm{Re}\,(e^{\mathrm{i}t}A)$, and $\mathrm{Im}A$ are compact. Finally, according to [70, Satz VI.3.4], $|\mathrm{Re}\,(e^{\mathrm{i}t}A)|$ and $A_\sharp^{1/2}$ are compact and self-adjoint. In Section 6.5, we will summarize properties of positive compact operators on H (to which $A_\sharp$ belongs). It will turn out in a moment that Theorem 6.3 allows us in particular to solve the acoustic as well as the electromagnetic inverse problem for a decent class of media. For almost everywhere absorbing media, we will apply the following theorem, which is significantly easier to prove than Theorem 6.3.

Theorem 6.4. *Let* H_1 *and* H_2 *be Hilbert spaces and* $A : H_1 \to H_1$, $B : H_1 \to H_2$, *and* $C : H_2 \to H_2$ *be linear and bounded operators such that*

$$A = B^* C B.$$

In addition, let there hold:

(i) B *is compact and injective.*

(ii) $\mathrm{Im}C$ *is coercive.*

Then $A_\sharp = \mathrm{Im}A$ *is positive, and the ranges of* B^* *and* $A_\sharp^{1/2}$ *coincide.*

Theorem 6.4 is a simple application of Theorem 4.1 from [41], with $\sigma = e^{\mathrm{i}3/2\pi} = -\mathrm{i}$. We note that under the conditions of Theorem 6.4, for $t = \frac{3}{2}\pi$ the conclusions of Theorems 6.3 and 6.4 are equivalent, with $A_\sharp = |\mathrm{Im}A| + \mathrm{Im}A = 2\,\mathrm{Im}A$ in the former. Hence, Theorem 6.4 can be considered as a corollary to Theorem 6.3. The next two sections give evidence that we can apply both theorems to our inverse problems. To this end, let for any $z \in \Pi$ the *probe functions* $\psi_{\alpha,z} \in L^2(\Gamma_s)$ and $\widetilde{\psi}_{\alpha,z} \in L^2(\Gamma_s, \mathbb{C}^3)$ be defined as in Theorems 6.1 and 6.2, respectively, i.e.

$$\psi_{\alpha,z}(x) = G_\alpha(z,x) \qquad \text{and} \qquad \widetilde{\psi}_{\alpha,z}(x) = k_0^2\, \mathfrak{G}_\alpha(z,x)\, p, \qquad x \in \Gamma_s, \qquad (6.6)$$

with some fixed $p \in \mathbb{R}^3$. We start with the treatment of the easier situation where the scattering medium is absorbing.

6.3 Absorbing media

We deal with absorbing media with a refraction index n (acoustic scattering) and a relative permittivity ε_r (electromagnetic scattering) such that the contrast $q = n - 1$ and $q = 1 - 1/\varepsilon_r$, respectively, satisfies

$$\operatorname{Im} q \geq c_1 |q| \qquad \text{almost everywhere in } \Omega \tag{6.7}$$

with some constant $c_1 > 0$.

6.3.1 The acoustic case

The following proposition ensures that under the condition (6.7) we can apply Theorem 6.4 to the acoustic near field operator M.

Proposition 6.5. *Using the notation from Chapter 3, there hold:*

- *$H_1 = L^2(\Gamma_s)$ and $H_2 = L^2(\Omega)$ are Hilbert spaces.*
- *The operators $A = M$, $B = H_{\Gamma_s}$, and $C = T$ are bounded linear operators.*
- *The operator $B = H_{\Gamma_s}$ fulfills Theorem 6.4 (i) by Proposition 3.5.*
- *Given (6.7), the operator $C = T$ fulfills Theorem 6.4 (ii) by Theorem 3.8 (iii).*

Finally, by the combination of Theorems 6.1 and 6.4, we gain the following characterization of the scattering medium.

Theorem 6.6. *The point $z \in \Pi$ lies in Ω if and only if $\psi_{\alpha,z}$ from (6.6) lies in the range of $(\operatorname{Im} M)^{1/2}$.*

In practical applications, the operator M is computed approximately from the physical near field operator $\widetilde{M}$ using the relation (5.8).

6.3.2 The electromagnetic case

Theorem 6.4 can also be applied to the electromagnetic near field operator $\mathcal{M}$. The next proposition collects the necessary results.

Proposition 6.7. *Using the notation from Chapter 4, there hold:*

- *$H_1 = L^2(\Gamma_s, \mathbb{C}^3)$ and $H_2 = L^2(\Omega, \mathbb{C}^3)$ are Hilbert spaces.*
- *The operators $A = \mathcal{M}$, $B = \mathcal{H}_{\Gamma_s}$, and $C = \mathcal{T}$ are bounded linear operators.*

- *The operator $B = \mathcal{H}_{\Gamma_s}$ fulfills Theorem 6.4 (i) by Proposition 4.5.*
- *Given (6.7), the operator $C = \mathcal{T}$ fulfills Theorem 6.4 (ii) by Theorem 4.9 (iii).*

Combining Theorems 6.2 and 6.4 yields

Theorem 6.8. *The point $z \in \Pi$ lies in Ω if and only if $\widetilde{\psi}_{\alpha,z}$ from (6.6) lies in the range of $(\operatorname{Im}\mathcal{M})^{1/2}$.*

In practice, the operator $\mathcal{M}$ can be approximated by means of the physical near field operator $\widetilde{\mathcal{M}}$ and the relation (5.23).

6.4 More general media

Supplementing Section 6.3, we now consider non-absorbing media and media which might be only partially absorbing. The conditions which we impose on the contrast q here differ in the cases of acoustic scattering and electromagnetic scattering.

6.4.1 The acoustic case

Assume that the contrast $q = n - 1$ obeys

$$\operatorname{Re}(e^{\mathrm{i}t}q) \geq c_0|q| \qquad \text{almost everywhere in } \Omega \tag{6.8}$$

with some constants $t \in [0, 2\pi)$ and $c_0 > 0$. Then we can make use of Theorem 6.3, as affirmed next.

Proposition 6.9. *Using the notation from Chapter 3, there hold:*

- *$X = X^* = U = L^2(\Omega)$ is a Hilbert space and as such a reflexive Banach space. $H = L^2(\Gamma_s)$ is a Hilbert space.*
- *The operators $A = M$, $B = H_{\Gamma_s}$, and $C = T$ are bounded linear operators.*
- *The operator $B = H_{\Gamma_s}$ fulfills Theorem 6.3 (i) by Proposition 3.5.*
- *Given (6.8), the operator $C = T$ fulfills Theorem 6.3 (ii), (iii), and (iv)-a by Theorem 3.8 (i), (ii), and (iv).*

Thus, via Theorem 6.1 we obtain the following.

Theorem 6.10. *The point $z \in \Pi$ lies in Ω if and only if $\psi_{\alpha,z}$ from (6.6) lies in the range of $M_\sharp^{1/2}$ where $M_\sharp = |\operatorname{Re}(e^{\mathrm{i}t}M)| + \operatorname{Im} M$.*

6.4.2 The electromagnetic case

Let the contrast $q = 1 - 1/\varepsilon_r$ be real-valued and satisfy

$$q > 0 \quad \Longleftrightarrow \quad \varepsilon_r > 1 \qquad \text{almost everywhere in } \Omega. \tag{6.9}$$

Proposition 6.11 verifies that Theorem 6.3 applies then with $t = 0$.

Proposition 6.11. *Using the notation from Chapter 4, there hold:*

- $X = X^* = U = L^2(\Omega, \mathbb{C}^3)$ *and* $H = L^2(\Gamma_s, \mathbb{C}^3)$ *are Hilbert spaces.*
- *The operators* $A = \mathcal{M}$, $B = \mathcal{H}_{\Gamma_s}$, *and* $C = \mathcal{T}$ *are bounded linear operators.*
- *The operator* $B = \mathcal{H}_{\Gamma_s}$ *fulfills Theorem 6.3 (i) by Proposition 4.5.*
- *Given (6.9), the operator* $C = \mathcal{T}$ *fulfills Theorem 6.3 (ii) with* $t = 0$*, (iii), and (iv)-a by Theorem 4.9 (i), (ii), and (iv).*

Using this and Theorem 6.2, we arrive at

Theorem 6.12. *The point* $z \in \Pi$ *lies in* Ω *if and only if* $\widetilde{\psi}_{\alpha,z}$ *from (6.6) lies in the range of* $\mathcal{M}_\sharp^{1/2}$ *where* $\mathcal{M}_\sharp = |\mathrm{Re}\,\mathcal{M}| + \mathrm{Im}\,\mathcal{M}$.

6.5 Regularization of the Factorization Method

In the statements of Theorems 6.6, 6.8, 6.10, and 6.12 we refer to the ranges of operators which depend on the exact artificial near field operator M (acoustic case) or $\mathcal{M}$ (electromagnetic case). However, all we can compute from given data are approximants to M and $\mathcal{M}$, respectively. It is not yet clear in which sense, if at all, the characterizations of the medium Ω in the previous sections can be met "in the limit" when one has to deal with a sequence of approximants to the artificial near field operator. Since the same questions arise for the acoustic and the electromagnetic case, we address only the acoustic case in the following discussion. To state the problem more precisely: How can we benefit from the range criterions in Theorems 6.6 and 6.10 to recover Ω (asymptotically) when we are given a sequence of operators M_j, $j \in \mathbb{N}$, which satisfy $\|M_j - M\| \to 0$ in the operator norm for $j \to \infty$? In this sort of abstract question, we leave aside the difficulty to obtain approximants M_j from measurement data in the form of $\widetilde{MP}_{\delta_j}$, defined in Theorem 5.4, with $\delta_j \to 0$ for $j \to \infty$. To tackle this question, we first note that the operators $\mathrm{Im}\,M$ in Theorem 6.6 and $M_\sharp$ in Theorem 6.10 are positive and compact.

To collect some basic properties of such operators, let $F \in \mathcal{L}(H)$ denote a positive and compact operator on a complex Hilbert space H. Then F is self-adjoint [60, Theorem 12.32] and injective. Due to the compactness, the spectrum of F is the union of $\{0\}$ and the at most countable set $\{\lambda_i\}$ of eigenvalues of F. The eigenvalues $\lambda_i \neq 0$ have finite geometric multiplicities $d_i = \dim\ker(\lambda_i \,\mathrm{id} - F)$ and accumulate at most at zero, see Theorems 4.18 and 4.25 in [60]. The self-adjointness of F implies that all λ_i are real, the eigenvectors belonging to different eigenvalues are orthogonal, and the algebraic multiplicities $a_i = \dim N(\lambda_i)$, $N(\lambda_i) = \{x \in H : \exists k \geq 1 \ (\lambda_i \,\mathrm{id} - F)^k x = 0\}$, for $\lambda_i \neq 0$ coincide with the geometric ones, cf. Lemma VI.3.1 and p. 299f. in [70]. The spectral theorem [70, Theorem VI.3.2] states that in fact there is an orthonormal basis of H which consists of eigenvectors of F. Due to the injectivity of F and the geometric multiplicities of the non-zero eigenvalues being finite, F has infinitely many eigenvalues if H is infinite-dimensional. Since F is positive, all its eigenvalues are indeed positive. Now, let (σ_n, ϕ_n) denote a sequence of pairs of eigenvalues and normalized eigenvectors of F, where σ_n takes the value of each λ_i according to its multiplicity, and all ϕ_n which belong to those σ_n equal to λ_i span the eigenspace $\ker(\lambda_i \,\mathrm{id} - F)$. Then F can be represented by

$$Fx = \sum_n \sigma_n \langle x, \phi_n \rangle \phi_n \qquad \text{for all } x \in H.$$

As a positive operator, F has a unique square root $F^{1/2} \in \mathcal{L}(H)$ [60, Theorem 12.33]. In fact, by [70, Satz VI.3.4], the spectral representation of $F^{1/2}$, and the positivity of F, the operator $F^{1/2}$ is seen to be positive and compact. Hence, it shares all the above properties of F.

Now, let, for instance, $\{(\sigma_n, \phi_n)\}$ denote an eigensystem of $(\mathrm{Im}\,M)^{1/2}$. We assume that the σ_n are ordered such that $\sigma_1 \geq \sigma_2 \geq \ldots > 0$. Theorem 6.6 asserts that a point $z \in \Pi$ lies in Ω if and only if the equation

$$(\mathrm{Im}\,M)^{1/2} g = \psi_{\alpha,z} \tag{6.10}$$

has a solution. Equivalently, $z \in \Pi$ lies in Ω if and only if the Picard sequence

$$p_z(N) = \sum_{n=1}^{N} \frac{|\langle \psi_{\alpha,z}, \phi_n \rangle|^2}{\sigma_n^2}, \qquad N \in \mathbb{N}, \tag{6.11}$$

stays bounded for $N \to \infty$. This demonstrates that the identification of Ω by the operator $(\mathrm{Im}\,M)^{1/2}$ is an ill-posed problem since σ_n decay to zero as n goes to infinity, cp. Section 4.3 in [17]. It is an unconventional ill-posed inverse problem in the sense that the 'data' are given by the specified function $\psi_{\alpha,z} = G_\alpha(z,\cdot)$ and that the focus is not on the computation of an approximate solution to (6.10), but on the solvability of (6.10) itself. The ill-posedness comes into play with the numerical representations of $\psi_{\alpha,z}$ and of the eigensystem $\{(\sigma_n,\phi_n)\}$ of $(\mathrm{Im}\,M)^{1/2}$. This in turn connects the nature of the identification problem with the type of our initial question. Since we only have a sequence of noisy operators M_j at hand, the challenge is to imitate for $j\to\infty$ the behavior of the exact Picard sequence $p_z(N)$ for $N\to\infty$, using the perturbed eigensystem $\{(\sigma_n^{(j)},\phi_n^{(j)})\}$ of $(\mathrm{Im}\,M_j)^{1/2}$. Aside, we remark that the sequence $\sigma_n^{(j)}$, $n\in\mathbb{N}$, could be finite, i.e. $\sigma_n^{(j)}=0$ for big enough n. This problem has been solved for a related setting by LECHLEITER in [48], see also Section I-6 in [49]. His method provides a suitable truncation of the perturbed Picard sequence. We want to address shortly an important ingredient for that. To this end, let A denote the operator of interest in [48] and $\{A_j\}$ a family of approximants which satisfy $\|A-A_j\|\to 0$ for $j\to\infty$. The exact Picard sequence therein refers to the square root of the operator $A_\sharp = |\mathrm{Re}\,A| + \mathrm{Im}\,A$. Preceding the construction of the truncation index, LECHLEITER uses an estimate from [66] to conclude from the norm convergence of A_j to A that

$$\||\mathrm{Re}\,A_j| - |\mathrm{Re}\,A|\| \to 0 \quad\text{and}\quad \|(A_j)_\sharp - A_\sharp\| \to 0 \qquad \text{for } j\to\infty,$$

where $(A_j)_\sharp$ is defined analogously to $A_\sharp$. To adapt the method to our setting, we have to guarantee that $\|\mathrm{Im}\,M_j - \mathrm{Im}\,M\| \to 0$ in the case of absorption (6.7) and that $\|(M_j)_\sharp - M_\sharp\| \to 0$ with $M_\sharp = |\mathrm{Re}\,(e^{\mathrm{i}t}M)| + \mathrm{Im}\,M$ in the more general case (6.8). Based on Theorem 5.4 and $M_j = \widetilde{M}P_{\delta_j}$, the former convergence is a direct consequence of

$$\|\mathrm{Im}\,M_j - \mathrm{Im}\,M\| \le \frac{1}{2}\|M_j + M_j^* - (M+M^*)\| \le \|M_j - M\|,$$

while the latter one can be shown using again the estimate from [66]. Operator estimates of this type are discussed thoroughly in [52]. Let now $p_z^{(j)}(N)$, $N\in\mathbb{N}$, denote the Picard sequence with respect to the eigensystem $\{(\sigma_n^{(j)},\phi_n^{(j)})\}$ of either $(\mathrm{Im}\,M_j)^{1/2}$ or $((M_j)_\sharp)^{1/2}$, depending on whether the contrast q obeys (6.7) or (6.8). With our preparation, a mapping $j\mapsto N(j)$ can be designed following [48] such that $j\mapsto p_z^{(j)}(N(j))$ stays bounded if and only if $z\in\Omega$. The incorporation of this regularized Picard criterion into Theorems 6.6 and 6.10 then yields an asymptotic identification of the medium by known data only. Similar statements hold for the electromagnetic inverse problem, substituting M by $\mathcal{M}$ and M_j by $\mathcal{M}_j = \widetilde{\mathcal{M}}\mathcal{P}_{\delta_j}$, cf. Theorem 5.6. We finally note that the proposed Picard technique requires the knowledge of a growing part of the disturbed eigensystem $\{(\sigma_n^{(j)}, \phi_n^{(j)})\}$ for $j \to \infty$, which might become expensive in practice. However, first considerations let us suppose it to be very difficult to construct an alternative regularization method for the range criterions which at least avoids the knowledge of the eigenfunctions.

Chapter 7

Numerical solvers

In this final chapter, we develop and apply numerical solvers for the direct and the inverse scattering problem which we investigated in theory in the previous chapters. We restrict ourselves here to the acoustic case in 2D, but remark that a big part of the presentation can be carried over in a straightforward way to the 3D case and also to the electromagnetic setting. In the first section, we adapt to our problem a fast solver for the Lippmann-Schwinger equation which has been proposed by VAINIKKO in [68] and discussed in detail in the monograph [61]. In particular, we demonstrate features of this approach which arise from the α-quasi-periodicity of our setting. The original and the adapted solver work properly only for continuous contrasts (we show a numerical example for this later). However, the optical devices which are produced up to now and which are of interest in today's applications have discontinuous contrasts. A main drawback of the continuity requirement is that it implies a smooth transition of the contrast to zero at the boundary of the medium. Thus, the usability of the solver for exact practical computations is limited. At the end of the first section, we propose a variant of the solver for our Lippmann-Schwinger equation which can treat contrasts which are piecewise constant on rectangles. This class of contrasts is covered by our theoretical considerations in the previous chapters and includes a major part of the devices used today. On the downside, our current implementation of the new solver requires a very expensive one-time precomputation for each medium geometry. This, however, can be improved significantly, we will comment on this. We show some numerical results for the adapted as well as the new method. The second section deals with the inverse problem of reconstructing the shape of the (inhomogeneous) medium from scattering data. Our intention is a numerical validation of our variant of the Factorization Method as a solver for this problem. The computations rely on the application of the solvers from the first section for the simulation of the direct problem and the setup of the numerical near field operator used in the implementation of the Factorization Method. We describe the overall computation scheme to show how the main results of this thesis are integrated in a practical realization. Afterwards, we present numerical examples for a smooth and a piecewise constant contrast and illustrate the dependence of the reconstructions on the different parameters. We close the thesis with a short conclusion.

7.1 Direct problem: A fast solver for the α-quasi-periodic Lippmann-Schwinger equation

7.1.1 The $\star$-periodic Lippmann-Schwinger equation

We start by rephrasing the α-quasi-periodic acoustic Lippmann-Schwinger equation as

$$u_\alpha(x) = u^i_\alpha(x) + k_0^2 \int_\Omega G_\alpha(y,x)\, q(y)\, u_\alpha(y)\,\mathrm{d}y, \qquad x \in \Pi. \qquad (\alpha\text{-LSE})$$

The integral is over $\Omega = \Omega' \cap \Pi$, where $\overline{\Omega'} \supseteq \operatorname{supp} q$, the contrast $q = n-1$ is assumed to be 2π-periodic in the x_1-dimension, and $\Pi = (-\pi,\pi)\times\mathbb{R}$ is the 2D unit cell. Moreover, $\Gamma_i \subset \Omega^{\mathrm{ext}} = \Pi\backslash\overline{\Omega}$ consists of horizontal lines and G_α denotes the α-quasi-periodic scalar Green's function for the 2D Helmholtz equation in free field conditions. It is given by

$$G_\alpha(y,x) = \frac{\mathrm{i}}{4\pi}\sum_{z\in Z}\frac{1}{\beta_z} e^{\mathrm{i}(\alpha_z\cdot(x-y)+\beta_z|x_2-y_2|)} \qquad (7.1)$$

for $x,y \in \Pi$ with $x_2 \neq y_2$, where $\alpha \in \mathbb{R}\times\{0\}$, $Z = \mathbb{Z}\times\{0\}$, $\alpha_z = \alpha + z$, and $\beta_z = \sqrt{k_0^2 - |\alpha_z|^2} \neq 0$ for all $z \in Z$. Point sources on Γ_i generate the incident acoustic field u^i_α. Again, u_α is the total field, i.e. the sum of u^i_α and the associated scattered field u^s_α. The difference between the acoustic Lippmann-Schwinger equation (3.37) in Chapter 3 and equation (α-LSE) above is that (3.37) is formulated for the scattered field rather than the total field, with a generalized incidence and in three dimensions. To emphasize that the Green's function G_α depends only on the difference of its arguments, we change the notation for this chapter from $G_\alpha(y,x)$ to $G_\alpha(x-y)$. Moreover, we write f_α for u^i_α. Recall that we identify any periodic or α-quasi-periodic function with its restriction to the unit cell Π. Now, we define the set

$$C_r = \{x \in \Pi : |x_2| < r\} \qquad (7.2)$$

with $r > 0$ such that $\overline{\Omega} \cap \Pi \subset C_r$ and $\overline{C_r} \cap \Gamma_i = \emptyset$. Since the periodic contrast $q : \Pi \to \mathbb{C}$ is supported in $\overline{\Omega}$, we may enlarge without effect the integration domain in (α-LSE) to C_r. Thus, we obtain the equation

$$u_\alpha(x) = f_\alpha(x) + k_0^2 \int_{C_r} G_\alpha(x-y)\, q(y)\, u_\alpha(y)\, \mathrm{d}y, \qquad x \in \Pi. \tag{7.3}$$

Obviously, the function u_α is completely determined by its restriction to the integration domain C_r and even to Ω. It is now important to note that $x, y \in C_r$ implies $x - y \in 2 \cdot C_r$ and that the y-dependent part of the integrand in (7.3) is periodic in the first coordinate, since $G_\alpha(x - \cdot)$ is $-\alpha$-quasi-periodic and $q u_\alpha$ is α-quasi-periodic. The latter function might, however, not have a well-defined trace on $\partial\Pi$, depending on the regularity of q. In consequence, only the restriction of G_α to the set $C_{2r} = (2 \cdot C_r) \cap \Pi$ counts for the integral in (7.3). This is a simple implication of the α-quasi-periodicity of our problem. VAINIKKO'S solver (see Section 10.5 in [61]) is based on the observation that a modification of the integral kernel in the Lippmann-Schwinger equation outside a specific 'relevant' region does not affect the computation of the total field. Stated precisely for our setting, a modification of G_α in $\Pi \backslash \overline{C_{2r}}$ does not change the function u_α in C_r. Therefore, we introduce a third set $C_{\tilde{r}}$ with $\tilde{r} \geq 2r$ such that $C_{2r} \subseteq C_{\tilde{r}}$ and a new kernel K_α on $\overline{C_{\tilde{r}}}$ by

$$K_\alpha(y) = \begin{cases} G_\alpha(y) & , y \in \overline{C_{2r}} \\ 0 & , y \in \overline{C_{\tilde{r}}} \backslash \overline{C_{2r}} \end{cases}. \tag{7.4}$$

Here, we include a part of the boundary of the unit cell for consistency of the following presentation. To avoid an overload of notation, we redefine

$$f_\alpha(y) = \begin{cases} f_\alpha(y) & , y \in \overline{C_{2r}} \\ 0 & , y \in \overline{C_{\tilde{r}}} \backslash \overline{C_{2r}} \end{cases}$$

and denote $u_\alpha|_{\overline{C_{\tilde{r}}}}$ now by u_α. Then, we extend K_α, f_α, and u_α from $\overline{C_{\tilde{r}}}$ to $\mathbb{R}^2$ as functions which are α_1-quasi-periodic in the first coordinate and $2\tilde{r}$-periodic in the second coordinate. We call such functions $\star$-*periodic* for short. In addition, we extend the contrast q from $\overline{C_{\tilde{r}}}$ to $\mathbb{R}^2$ as a $(2\pi, 2\tilde{r})$-periodic function. This preserves the form of the original contrast q in the strip $\{x \in \mathbb{R}^2 : |x_2| < \tilde{r}\}$, but comes with a new x_2-dependence below and above this strip. For convenience, we use the same symbols for the extended functions as for their restrictions to $\overline{C_{\tilde{r}}}$. We point out that the new kernel K_α differs from the Green's function G_α only in the dependence on the second coordinate. Enlarging again the integration domain, we arrive at the $\star$-periodic equation

$$u_\alpha(x) = f_\alpha(x) + k_0^2 \int_{C_{\tilde{r}}} K_\alpha(x-y)\, q(y)\, u_\alpha(y)\, \mathrm{d}y, \qquad x \in \overline{C_{\tilde{r}}}. \tag{7.5}$$

In this equation, the characteristic part of all functions lies in $C_{\tilde{r}}$ and the integral is over $C_{\tilde{r}}$. This is the starting point for an efficient numerical solution of (7.5) by a Fourier technique. We finish our preparations and denote by $H_\star^\mu(C_{\tilde{r}})$ the space of $\star$-periodic functions of H^μ-regularity, with μ chosen such that they have a well-defined trace on $\partial C_{\tilde{r}}$. In fact, from now on we assume

$$q \in H_{\text{per}}^\mu(C_{\tilde{r}}) \qquad \text{and} \qquad q f_\alpha \in H_\star^\mu(C_{\tilde{r}}) \qquad \text{for some } \mu > 1, \tag{7.6}$$

where $H_{\text{per}}^\mu(\cdot)$ stands for $H_{\alpha=0}^\mu(\cdot)$. According to Sobolev's Lemma (cf. [69]), q and $q f_\alpha$ are then continuous in $C_{\tilde{r}}$. Moreover, they have continuous extensions to $\partial C_{\tilde{r}}$. Multiplication of (7.5) with q finally yields

$$w_\alpha(x) = (q f_\alpha)(x) + k_0^2 q(x) \int_{C_{\tilde{r}}} K_\alpha(x-y)\, w_\alpha(y)\, \mathrm{d}y, \qquad x \in \overline{C_{\tilde{r}}}, \tag{$\star$-LSE}$$

with $w_\alpha = q u_\alpha$. This is the *$\star$-periodic Lippmann-Schwinger equation* which we consider in the following. Simple considerations make clear that the equations (7.3), (7.5), and ($\star$-LSE) are equivalent with respect to existence and uniqueness of a solution. For instance, if (7.5) has at most one solution, then ($\star$-LSE) has at most one solution. To show this, assume on the contrary to the conclusion that $w_{\alpha,1}$ and $w_{\alpha,2}$ solve ($\star$-LSE) and $w_{\alpha,1} \not\equiv w_{\alpha,2}$. Then by ($\star$-LSE) there also holds $w_{\alpha,1}|_{\operatorname{supp} q \cap \Pi} \not\equiv w_{\alpha,2}|_{\operatorname{supp} q \cap \Pi}$. But then $((w_{\alpha,1} - w_{\alpha,2})/q)|_{\operatorname{supp} q \cap \Pi} \not\equiv 0$ solves (7.5) in $\operatorname{supp} q \cap \Pi$ with $f_\alpha \equiv 0$ and determines (uniquely) a nontrivial solution to (7.5) in $\overline{C_{\tilde{r}}}$ with $f_\alpha \equiv 0$. This is a contradiction to the assumption that (7.5) has at most one solution. If u_α is a solution to (7.5), then (the $\star$-periodic extension of) $q u_\alpha$ solves ($\star$-LSE).

7.1.2 Trigonometric collocation

The equation ($\star$-LSE) is a Fredholm volume integral equation of the second kind. In order to solve it numerically, we set up a finite-dimensional collocation problem for ($\star$-LSE). To start, we introduce the index set

$$\mathbb{Z}_N^2 = \left\{ j \in \mathbb{Z}^2 : -\frac{N}{2} < j_k \leq \frac{N}{2},\, k = 1,2 \right\}, \qquad N \in 2 \cdot \mathbb{N}, \tag{7.7}$$

and the N^2-dimensional linear space $T_{\star,N}$ of $\star$-periodic trigonometric polynomials on $\overline{C_{\tilde{r}}}$ of the form $v_{\alpha,N} = \sum_{j\in\mathbb{Z}_N^2} a_j \varphi_{\alpha,j}$ with $a_j \in \mathbb{C}$ and

$$\varphi_{\alpha,j}(x) = (4\pi\tilde{r})^{-1/2} e^{\mathrm{i}\,\alpha\cdot x} e^{\mathrm{i}\pi(j_1x_1/\pi + j_2x_2/\tilde{r})}, \qquad x \in \overline{C_{\tilde{r}}}.$$

We note that $\varphi_{\alpha,j}$, $j \in \mathbb{Z}^2$, form an orthonormal basis of $L^2(C_{\tilde{r}})$. We will use the abbreviations

$$(j./c)\cdot x = j_1x_1/\pi + j_2x_2/\tilde{r}, \qquad c = (\pi,\tilde{r}), \qquad \widetilde{c} = (4\pi\tilde{r})^{-1/2},$$

so that $\varphi_{\alpha,j}(x) = \widetilde{c}\, e^{\mathrm{i}(\alpha+\pi j./c)\cdot x}$. Associated with the space $T_{\star,N}$, we define the interpolation projection $Q_N : H^{\mu}_{\star}(C_{\tilde{r}}) \to T_{\star,N}$ for $\mu > 1$ by claiming

$$Q_N v_\alpha \in T_{\star,N}, \quad (Q_N v_\alpha)(j\odot h_N) = v_\alpha(j\odot h_N) \quad \text{for all } j \in \mathbb{Z}_N^2, \quad h_N = \frac{2c}{N}, \tag{7.8}$$

for $v_\alpha \in H^{\mu}_{\star}(C_{\tilde{r}})$, where $\odot$ stands for a componentwise multiplication. This means that Q_N performs a collocation on the grid $\mathcal{G}_N = \mathbb{Z}_N^2 \odot h_N$ in the space $T_{\star,N}$. The grid $\mathcal{G}_N$ lacks grid points on the left and at the bottom of $\overline{C_{\tilde{r}}}$, which is due to the $\star$-periodicity of the functions to be handled. The next theorem makes a statement about the approximation quality of Q_N in a certain Sobolev range, cp. Theorem 8.5.3 in [61].

Theorem 7.1. *For any $v_\alpha \in H^{\mu}_{\star}(C_{\tilde{r}})$ with $\mu > 1$ there holds*

$$\|v_\alpha - Q_N v_\alpha\|_\lambda \le c_{\lambda,\mu} N^{\lambda-\mu} \|v_\alpha\|_\mu \tag{7.9}$$

for any $0 \le \lambda \le \mu$, where $c_{\lambda,\mu} = 2^{\mu-\lambda}\left(2^\lambda \sum_{l,l'=0}^{\infty} (l^2+l'^2)^{-\mu}\right)^{1/2}$ and $\|\cdot\|_\mu$ denotes the norm of $H^\mu(C_{\tilde{r}})$.

We consider the approximate solution of ($\star$-LSE) by the trigonometric collocation method

$$w_{\alpha,N} = Q_N(q f_\alpha) + Q_N(k_0^2 q \mathcal{K}_\alpha w_{\alpha,N}) \tag{7.10}$$

where $\mathcal{K}_\alpha : L^2(C_{\tilde{r}}) \to H^2_{\star}(C_{\tilde{r}})$ is the integral operator given by

$$(\mathcal{K}_\alpha v)(x) = \int_{C_{\tilde{r}}} K_\alpha(x-y)\, v(y)\, \mathrm{d}y, \qquad x \in \overline{C_{\tilde{r}}}. \tag{7.11}$$

With $\mathcal{K}_\alpha$ being a convolution operator, one shows by means of the convolution theorem that

$$\mathcal{K}_\alpha \varphi_{\alpha,j} = \widetilde{c}^{\,-1}\widehat{K_\alpha}(j)\,\varphi_{\alpha,j}, \qquad j \in \mathbb{Z}^2, \tag{7.12}$$

holds, where $\widehat{K_\alpha}(j)$ are the Fourier coefficients of K_α with respect to $\{\varphi_{\alpha,j}\}$, $j \in \mathbb{Z}^2$. We derive a simple expression for these coefficients at the end of this subsection. The relation (7.12) will allow us to avoid any explicit integration later on in our numerical scheme. To this end, we also need to compute the Fourier coefficients $\widehat{v_{\alpha,N}}(j)$ for $j \in \mathbb{Z}_N^2$ of a function $v_{\alpha,N} \in T_{\star,N}$. However, these coefficients are simply given by

$$\begin{aligned} \widehat{v_{\alpha,N}}(j) &= \int_{C_{\tilde{r}}} v_{\alpha,N}(y)\,\overline{\varphi_{\alpha,j}(y)}\,\mathrm{d}y \\ &= \tilde{c}^{-1} N^{-2} \sum_{l \in \mathbb{Z}_N^2} v_{\alpha,N}(l \odot h_N)\, e^{-\mathrm{i}\alpha\cdot(l\odot h_N)}\, e^{-\mathrm{i}2\pi l\cdot j/N}, \qquad j \in \mathbb{Z}_N^2. \end{aligned} \tag{7.13}$$

The right-hand side of (7.13) is a discrete 2D Fourier transform of the modified node values $v_{\alpha,N}(j \odot h_N)\, e^{-\mathrm{i}\alpha\cdot(j\odot h_N)}$, $j \in \mathbb{Z}_N^2$. We abbreviate this transform by $\mathcal{F}_{\alpha,N}$. Complementary, $\mathcal{F}_{\alpha,N}^{-1}$ denotes the inverse transformation, given by

$$v_{\alpha,N}(j \odot h_N) = (\mathcal{F}_{\alpha,N}^{-1} \widehat{v_{\alpha,N}})(j) = \tilde{c}\, e^{\mathrm{i}\alpha\cdot(j\odot h_N)} \sum_{l \in \mathbb{Z}_N^2} \widehat{v_{\alpha,N}}(l)\, e^{\mathrm{i}2\pi l\cdot j/N}, \qquad j \in \mathbb{Z}_N^2.$$

Now, we can write the collocation problem (7.10) in the discrete form

$$\underline{w_{\alpha,N}}_N = \underline{(q f_\alpha)}_N + k_0^2\, \underline{q}_N \odot \mathcal{F}_{\alpha,N}^{-1} \widehat{\mathcal{K}_{\alpha N}}\, \mathcal{F}_{\alpha,N} \underline{w_{\alpha,N}}_N, \tag{7.14}$$

where $\underline{\times}_N$ denotes the nodal values of $\times$ on the grid $\mathcal{G}_N$ and $\widehat{\mathcal{K}_{\alpha N}}$ represents the pointwise multiplication with the coefficients $\tilde{c}^{-1}\widehat{K_\alpha}(j)$, $j \in \mathbb{Z}_N^2$. In order to restate (7.14) as a linear equation system, we need to introduce the invertible operator $T_N : \mathbb{C}^{N\times N} \to \mathbb{C}^{N^2}$ which converts a matrix into a vector by concatenation of its columns. One should keep in mind that $\underline{\times}_N$ is matrix-valued. Using the operator T_N, for (7.14) we obtain the equivalent system

$$A_{N^2} T_N \underline{w_{\alpha,N}}_N = T_N \underline{g_\alpha}_N, \qquad A_{N^2} = I_{N^2} - k_0^2 \mathcal{H}_{N^2}, \qquad \underline{g_\alpha}_N = \underline{(q f_\alpha)}_N, \tag{7.15}$$

where I_{N^2} is the identity in $\mathbb{C}^{N^2\times N^2}$ and $\mathcal{H}_{N^2} \in \mathbb{C}^{N^2\times N^2}$ is the matrix representation of the linear operator $\underline{q}_N \odot \mathcal{F}_{\alpha,N}^{-1} \widehat{\mathcal{K}_{\alpha N}}\, \mathcal{F}_{\alpha,N} : \mathbb{C}^{N\times N} \to \mathbb{C}^{N\times N}$, which is consistent with the element ordering by T_N. The application of the Fourier transformation $\mathcal{F}_{\alpha,N}$ and its inverse $\mathcal{F}_{\alpha,N}^{-1}$ each costs N^4 multiplications and additions when they are computed in the straightforward way. As the alternative of choice, one should use the fast Fourier transform (FFT) and the fast inverse transform for this task and by that reduce the complexity to $\mathcal{O}(N^2 \log N)$ arithmetical operations. From the vast amount of literature on this algorithm, we only cite the seminal paper [18] by *Cooley and Tukey* and the reference [11]. To use the FFT for some $v_{\alpha,N} \in T_{\star,N}$, we have to eliminate the α-quasi-periodicity in the sample values $v_{\alpha,N}(j \odot h_N)$ by multiplication with $e^{-\mathrm{i}\alpha\cdot(j\odot h_N)}$, $j \in \mathbb{Z}_N^2$, and also to take care of the index as well as the phase shifts in the data since the two-dimensional FFT: $\mathbb{C}^{N\times N} \to \mathbb{C}^{N\times N}$ computes the expression

$$(\mathrm{FFT}\,\underline{v})_j = \sum_{l_1,l_2=0}^{N-1} \underline{v}_l e^{-\mathrm{i}2\pi l\cdot j/N}, \qquad l = (l_1,l_2),$$

for $j \in \widetilde{\mathbb{Z}}_N^2 = \{j \in \mathbb{Z}^2 : 0 \le j_k < N, k = 1,2\}$, compare (7.7). This is accomplished by applying twice the shift theorem, giving

$$(\mathcal{F}_{\alpha,N} v_{\alpha,N})_{j-s} = \mathrm{FFT}\left(\left\{\underline{v}_{l-s} e^{\mathrm{i}2\pi/N l\cdot s}\right\}_{l\in\widetilde{\mathbb{Z}}_N^2}\right)_j e^{\mathrm{i}2\pi/N(j-s)s}, \qquad j \in \widetilde{\mathbb{Z}}_N^2,$$

where $s = (N/2-1, N/2-1)$ and $\underline{v}_n = v_{\alpha,N}(n \odot h_N)e^{-\mathrm{i}\alpha\cdot(n\odot h_N)}$, $n \in \mathbb{Z}_N^2$. Nevertheless, these additional operations do not destroy the order $\mathcal{O}(N^2 \log N)$. In (7.14) and (7.15), the FFT evaluation can not be realized, since in both equations the argument $\underline{w_{\alpha,N}}_N$ is the unknown variable and in (7.15) the operators $\mathcal{F}_{\alpha,N}$ and $\mathcal{F}_{\alpha,N}^{-1}$ are hard-coded as matrices into $\mathcal{H}_{N^2}$. However, since the application of the operator A_{N^2} is cheap, it is convenient to solve the linear system (7.15) by some iteration method. There, A_{N^2} does not need to be represented in matrix form, and we can take advantage of the fast evaluations of $\mathcal{F}_{\alpha,N}$ and $\mathcal{F}_{\alpha,N}^{-1}$. A very efficient solution method, which has the same complexity $\mathcal{O}(N^2 \log N)$ as the fast Fourier transforms, is made up by *two-grid iterations*. These are the subject of Subsection 7.1.3. Instead of (7.15), we consider primarily its Fourier counterpart, reading

$$\widehat{A}_{N^2} T_N \widehat{w_{\alpha,N}} = T_N \widehat{g_{\alpha,N}}, \qquad \widehat{A}_{N^2} = I_{N^2} - k_0^2 \widehat{\mathcal{H}}_{N^2}, \qquad \widehat{g_{\alpha,N}} = \mathcal{F}_{\alpha,N}\big(\underline{q f_\alpha}\big)_N, \quad (7.16)$$

where $\widehat{\mathcal{H}}_{N^2}$ is the matrix representation of the operator $\mathcal{F}_{\alpha,N}\big(\underline{q}_N \odot \mathcal{F}_{\alpha,N}^{-1} \widehat{\mathcal{K}_{\alpha N}}\big)$. The central result about the collocation method is the following, cp. Theorem 10.5.1 in [61].

Theorem 7.2. *Assume that $q \in H_{per}^{\mu}(C_{\tilde{r}})$ and $q f_\alpha \in H_\star^{\mu}(C_{\tilde{r}})$ for some $\mu > 1$. Moreover, let the homogeneous problem corresponding to (7.3) with $f_\alpha \equiv 0$ have only the trivial solution. Then equation ($\star$-LSE) has a unique solution $w_\alpha \in H_\star^{\mu}(C_{\tilde{r}})$, the collocation equation (7.10) has a unique solution $w_{\alpha,N} \in T_{\star,N}$ for sufficiently large N, and*

$$\|w_{\alpha,N} - w_\alpha\|_\lambda \le c N^{\lambda-\mu} \|w_\alpha\|_\mu, \qquad 0 \le \lambda \le \mu,$$

with $c > 0$ independent of α and N.

Proof. By the boundedness of $\mathcal{K}_\alpha$ as a mapping from $L^2(C_{\tilde{r}})$ to $H^2_\star(C_{\tilde{r}})$ and the assumption $q \in H^\mu_{\text{per}}(C_{\tilde{r}})$ with $\mu > 1$, the operator $q\mathcal{K}_\alpha$ is bounded from $L^2(C_{\tilde{r}})$ to $H_\star^{\min(\mu,2)}(C_{\tilde{r}})$ and hence is compact as an operator in $L^2(C_{\tilde{r}})$. Since the homogeneous integral equation corresponding to ($\star$-LSE) with $f_\alpha \equiv 0$ has only the trivial solution, the operator $\text{id} - k_0^2 q\mathcal{K}_\alpha \in \mathcal{L}(L^2(C_{\tilde{r}}))$ is boundedly invertible by the Riesz theory, cf. Theorem 1.16 in [16]. By Theorem 7.1 with $\mu > 1$ and $\lambda = 0$, it is seen that

$$\|q\mathcal{K}_\alpha - Q_N(q\mathcal{K}_\alpha)\|_{\mathcal{L}(L^2(C_{\tilde{r}}))} \to 0 \qquad \text{for } N \to \infty.$$

Hence, the inverse to $\text{id} - Q_N(k_0^2 q\mathcal{K}_\alpha)$ in $\mathcal{L}(L^2(C_{\tilde{r}}))$ exists for sufficiently large N, say $N \geq N_0$, and by a standard perturbation result the norm of the inverse is uniformly bounded in N, i.e.

$$\|(\text{id} - Q_N(k_0^2 q\mathcal{K}_\alpha))^{-1}\|_{\mathcal{L}(L^2(C_{\tilde{r}}))} \leq c' \qquad \text{for all } N \geq N_0 \tag{7.17}$$

for some constant $c' > 0$. Moreover, by (7.10) and ($\star$-LSE), there hold the equalities

$$\begin{aligned}(\text{id} - Q_N(k_0^2 q\mathcal{K}_\alpha))(w_{\alpha,N} - w_\alpha) &= Q_N(q f_\alpha) - w_\alpha + Q_N(k_0^2 q\mathcal{K}_\alpha w_\alpha)\\ &= Q_N w_\alpha - w_\alpha.\end{aligned}$$

Combining this with (7.17), the assertion follows for $\lambda = 0$ by Theorem 7.1,

$$\|w_{\alpha,N} - w_\alpha\|_0 \leq c' c_{0,\mu} N^{-\mu} \|w_\alpha\|_\mu.$$

For the case $0 < \lambda \leq \mu$, we exploit that the orthogonal projection $P_N : H^\mu_\star(C_{\tilde{r}}) \to T_{\star,N}$ satisfies

$$\|w_\alpha - P_N w_\alpha\|_\lambda \leq \left(\frac{N}{2}\right)^{\lambda-\mu} \|w_\alpha\|_\mu, \qquad \lambda \leq \mu,$$

cf. Lemma 8.5.1 in [61]. Using also the inverse inequality

$$\|v_{\alpha,N}\|_\lambda \leq 2^{-\lambda/2} N^\lambda \|v_{\alpha,N}\|_0 \qquad \text{for } v_{\alpha,N} \in T_{\star,N},$$

see p. 319 in [61], we estimate

$$\begin{aligned}\|w_{\alpha,N} - w_\alpha\|_\lambda &\leq \|w_{\alpha,N} - P_N w_\alpha\|_\lambda + \|w_\alpha - P_N w_\alpha\|_\lambda\\ &\leq \left(\frac{N}{\sqrt{2}}\right)^\lambda (\|w_{\alpha,N} - w_\alpha\|_0 + \|w_\alpha - P_N w_\alpha\|_0) + \left(\frac{N}{2}\right)^{\lambda-\mu} \|w_\alpha\|_\mu\\ &\leq (2^{-\lambda/2} c' c_{0,\mu} + 2^{\mu-\lambda/2} + 2^{\mu-\lambda}) N^{\lambda-\mu} \|w_\alpha\|_\mu.\end{aligned}$$

This completes the proof. □

Fourier coefficients of K_α

Now, we compute the Fourier coefficients $\widehat{K_\alpha}(j)$, $j \in \mathbb{Z}^2$, of the kernel K_α, which are used for the application of the integral operator $\mathcal{K}_\alpha$ in the Fourier space according to the relation (7.12). First, we note that the Helmholtz operator applied to $\varphi_{\alpha,j}$ yields

$$(\Delta + k_0^2)\varphi_{\alpha,j} = (k_0^2 - |\alpha + \pi j./c|^2)\varphi_{\alpha,j}. \tag{7.18}$$

Setting $\lambda_{\alpha,j} = k_0^2 - |\alpha + \pi j./c|^2$, there obviously holds $\lambda_{\alpha,j} = \lambda_{-\alpha,-j} \in \mathbb{R}$, and we assume in the following that $\lambda_{\alpha,j} \neq 0$ for all $j \in \mathbb{Z}^2$. Then, the relation (7.18) can be used to apply the representation formula (3.20) and derive

$$\begin{aligned}
\widehat{K_\alpha}(j) &= \int_{C_{\tilde{r}}} K_\alpha(y)\,\overline{\varphi_{\alpha,j}(y)}\,\mathrm{d}y \\
&= \frac{1}{\lambda_{\alpha,j}} \int_{C_{2r}} G_\alpha(y)\,(\Delta + k_0^2)\overline{\varphi_{\alpha,j}(y)}\,\mathrm{d}y \\
&= \frac{1}{\lambda_{\alpha,j}}\Bigg(-\overline{\varphi_{\alpha,j}(0)} + \int_{\partial C_{2r}} \left(G_\alpha(y)\,\frac{\partial}{\partial \nu}\overline{\varphi_{\alpha,j}(y)} - \frac{\partial}{\partial \nu} G_\alpha(y)\,\overline{\varphi_{\alpha,j}(y)} \right) \mathrm{d}s(y) \Bigg).
\end{aligned}$$

The boundary ∂C_{2r} can be decomposed as $\partial C_{2r} = C_2^+ \cup C_2^- \cup V$ such that C_2^+ and C_2^- are horizontal lines with $\nu = \pm e_2$ on $C_2^\pm$, respectively, and V is a union of vertical lines. Since $G_\alpha(y)\,\overline{\varphi_{\alpha,j}(y)}$ is periodic, the contributions on V in the boundary integral in the above expression cancel out. Whereas for non-periodic problems, Vainikko's solver is based on artificial periodic extensions of the functions involved, we can exploit here the problem-specific periodicity of the functions. This is a particular feature of our variant of Vainikko's solver and leads to a simple expression for the coefficients $\widehat{K_\alpha}(j)$, $j \in \mathbb{Z}^2$. Computing explicitly the boundary integral over the remaining lines C_2^+ and C_2^-, we get

$$\begin{aligned}
\int_{C_2^\pm} \frac{\partial}{\partial \nu} G_\alpha(y)\,\overline{\varphi_{\alpha,j}(y)}\,\mathrm{d}s(y) &= -\frac{\widetilde{c}}{4\pi}\int_{-\pi}^{\pi} e^{-\mathrm{i}(\alpha+\pi j./c)\cdot y} \sum_{z\in Z} e^{\mathrm{i}(\alpha_z\cdot y + \beta_z 2r)}\,\mathrm{d}y_1 \\
&= -\frac{\widetilde{c}}{4\pi}\int_{-\pi}^{\pi} e^{-\mathrm{i}\pi j./c\cdot y} \sum_{z\in Z} e^{\mathrm{i}(z\cdot y + \beta_z 2r)}\,\mathrm{d}y_1 \\
&= -\frac{\widetilde{c}}{2} e^{\mathrm{i}(\beta_{(j_1,0)} \mp \pi j_2/\tilde{r})2r},
\end{aligned}$$

and, found in a similar fashion,

$$\int_{C_2^\pm} G_\alpha(y)\,\frac{\partial}{\partial \nu}\overline{\varphi_{\alpha,j}(y)}\,\mathrm{d}s(y) = \pm\frac{\widetilde{c}\,\pi\, j_2}{2\;\tilde{r}}\frac{1}{\beta_{(j_1,0)}} e^{\mathrm{i}(\beta_{(j_1,0)} \mp \pi j_2/\tilde{r})2r}.$$

Thus, there holds

$$\int_{C_2^\pm} \left(G_\alpha(y) \frac{\partial}{\partial \nu} \overline{\varphi_{\alpha,j}(y)} - \frac{\partial}{\partial \nu} G_\alpha(y) \overline{\varphi_{\alpha,j}(y)} \right) \mathrm{d}s(y) = \frac{\widetilde{c}}{2} \left(1 \pm \frac{\pi j_2}{\tilde{r}} \frac{1}{\beta_{(j_1,0)}} \right) e^{\mathrm{i}(\beta_{(j_1,0)} \mp \pi j_2/\tilde{r})2r}.$$

Summing up the integrals over C_2^+ and C_2^-, for the coefficients $\widehat{K_\alpha}(j)$, $j \in \mathbb{Z}^2$, we finally obtain

$$\widehat{K_\alpha}(j) = -\frac{\widetilde{c}}{\lambda_{\alpha,j}} \left(1 - e^{\mathrm{i}\beta_{(j_1,0)}2r} \left[\cos\left(\frac{\pi j_2}{\tilde{r}} 2r \right) + \mathrm{i} \frac{\pi j_2}{\tilde{r}} \frac{1}{\beta_{(j_1,0)}} \sin\left(\frac{\pi j_2}{\tilde{r}} 2r \right) \right] \right). \tag{7.19}$$

We remark that for $\tilde{r} = 2r$ this expression simplifies further to

$$\widehat{K_\alpha}(j) = -\frac{\widetilde{c}}{\lambda_{\alpha,j}} \left(1 - (-1)^{j_2} e^{\mathrm{i}\beta_{(j_1,0)}2r} \right), \qquad j \in \mathbb{Z}^2. \tag{7.20}$$

Due to the form of $\lambda_{\alpha,j}$ and the fact that $\beta_{(j_1,0)}$ becomes purely imaginary for sufficiently big modulus of j_1, there holds

$$\widehat{K_\alpha}(j) = \mathcal{O}(|j|^{-2}).$$

This implies that the convolution operator $\mathcal{K}_\alpha$ defined in (7.11) is bounded as a mapping from $L^2(C_{\tilde{r}})$ to $H^2_\star(C_{\tilde{r}})$.

7.1.3 Two-grid iteration scheme

In this subsection, we formulate a two-grid iteration scheme for the efficient solution of the collocation equation (7.10). Theorem 7.2 asserts that for sufficiently large N we obtain from (7.10) a suitable approximation to the solution to ($\star$-LSE). Our construction of the scheme follows Subsections 10.5.3 and 10.5.4 in [61], see also Section 3.7 in [68].

So, let $N \in \mathbb{N}$ be even and fixed. We define the function $g_{\alpha,N} \in T_{\star,N}$ and the operators $\mathcal{T}_M : T_{\star,N} \to T_{\star,M}$ for even $M \leq N$ by

$$g_{\alpha,N} = Q_N(q f_\alpha) \qquad \text{and} \qquad \mathcal{T}_M = Q_M(k_0^2 q \mathcal{K}_\alpha). \tag{7.21}$$

Then the collocation equation (7.10) turns into $(\mathrm{id} - \mathcal{T}_N) w_{\alpha,N} = g_{\alpha,N}$. Applying the operator $(\mathrm{id} - \mathcal{T}_M)^{-1}$ to both sides of this equation yields

$$w_{\alpha,N} = \mathcal{T}_{M,N} w_{\alpha,N} + g_{M,N} \tag{7.22}$$

where

$$\mathcal{T}_{M,N} = (\mathrm{id} - \mathcal{T}_M)^{-1}(\mathcal{T}_N - \mathcal{T}_M) \qquad \text{and} \qquad g_{M,N} = (\mathrm{id} - \mathcal{T}_M)^{-1} g_{\alpha,N}.$$

We recall that the operator $(\mathrm{id} - \mathcal{T}_M)^{-1}$ exists for sufficiently large M, according to the proof of Theorem 7.2. From Theorem 7.1 and estimate (7.17), we conclude that $\|\mathcal{T}_{M,N}\|_{\mathcal{L}(L^2(C_{\tilde{r}}))}$ is small, and so it is reasonable to apply the *two-grid iteration*

$$w_{\alpha,N}^{(j)} = \mathcal{T}_{M,N} w_{\alpha,N}^{(j-1)} + g_{M,N}, \qquad j = 1,2,\ldots, \tag{7.23}$$

starting e.g. from $w_{\alpha,N}^{(0)} = w_{\alpha,M} = (\mathrm{id} - \mathcal{T}_M)^{-1} g_M$. Concerning the approximation quality of the j-th iterate $w_{\alpha,N}^{(j)}$ and the choice of the stopping index $N_{\max} \geq j$, we refer to the analog discussion in Section 3.7 in [68]. With $(\mathrm{id} - \mathcal{T}_M)^{-1} = \mathrm{id} + (\mathrm{id} - \mathcal{T}_M)^{-1} \mathcal{T}_M$, the equation (7.23) can be rewritten as

$$w_{\alpha,N}^{(j)} = \left[\mathrm{id} + (\mathrm{id} - \mathcal{T}_M)^{-1} \mathcal{T}_M\right] \left[(\mathcal{T}_N - \mathcal{T}_M) w_{\alpha,N}^{(j-1)} + g_{\alpha,N}\right], \qquad j = 1,2,\ldots, \tag{7.24}$$

where the inverse $(\mathrm{id} - \mathcal{T}_M)^{-1}$ is applied only to functions in $T_{\star,M}$. This feature is the main factor for the performance of the two-grid iteration scheme. From now on, we assume that the coarsening factor $D = N/M$ is integer. Rather than using the representation (7.24) in the finite-dimensional function space $T_{\star,N}$, for the numerical computation it is appropriate to use its discrete Fourier form

$$\begin{aligned} \widehat{w_{\alpha,N}^{(j)}} = {} & \left[I_N + k_0^2 P_{N,M} T_M^{-1} \widehat{A}_{M^2}^{-1} T_M \mathcal{F}_{\alpha,M}\left(\underline{q}_M \odot R_{M,N} \mathcal{F}_{\alpha,N}^{-1} \widehat{\mathcal{K}_{\alpha N}}\right)\right] \cdot \\ & \cdot \left[k_0^2 \left(\mathcal{F}_{\alpha,N}(\underline{q}_N \odot) - P_{N,M} \mathcal{F}_{\alpha,M}(\underline{q}_M \odot R_{M,N})\right) \mathcal{F}_{\alpha,N}^{-1} \widehat{\mathcal{K}_{\alpha N}} \widehat{w_{\alpha,N}^{(j-1)}} + \widehat{g_{\alpha,N}}\right] \end{aligned} \tag{7.25}$$

with $j = 1,2,\ldots$, starting from $\widehat{w_{\alpha,N}^{(0)}} = \widehat{w_{\alpha,M}} = P_{N,M} T_M^{-1} \widehat{A}_{M^2}^{-1} T_M \widehat{g_{\alpha,M}}$. Here, $\widehat{A}_{M^2}$, $\widehat{\mathcal{K}_{\alpha N}}$, and T_M are the operators introduced in Subsection 7.1.2, and $P_{N,M}$ and $R_{M,N}$ are the prolongation and restriction operators defined by

$$(P_{N,M} \underline{v_M})(j) = \begin{cases} \underline{v_M}(j) & , j \in \mathbb{Z}_M^2 \\ 0 & , j \in \mathbb{Z}_N^2 \setminus \mathbb{Z}_M^2 \end{cases} \qquad \text{and} \qquad (R_{M,N} \underline{v_N}) = \underline{v_N}(J_{M,N}),$$

where

$$J_{M,N} = \left\{ \left(j - \frac{N}{2}, k - \frac{N}{2}\right) : j,k = \frac{N}{M}, \frac{2N}{M}, \ldots, N \right\}$$

and $\underline{v_N}(J_{M,N})$ is reindexed by $j \in \mathbb{Z}_M^2$.

We remark that a procedure for extending this two-grid scheme to an even more efficient multi-grid method is proposed in [34], Sections 4.3–4.5, therein formulated for an electromagnetic scattering problem in 3D.

7.1.4 Extension to discontinuous contrasts

The collocation method (7.10) for the approximate solution of ($\star$-LSE) yields a proper result only for sufficiently smooth contrasts (and appropriate values of N). We are now going to construct a related, yet new collocation method which can treat a class of discontinuous contrasts, precisely those which are piecewise constant on rectangles. This includes many contrasts given in current applications. Let us start with a short discussion of the arguments underlying the collocation method (7.10). The choice of the ansatz space $T_{\star,N}$ is motivated by the hope that the solution $w_{\alpha,N} \in T_{\star,N}$ to (7.10) approximates the solution $w_\alpha = q u_\alpha$ to ($\star$-LSE) arbitrarily well for increasing N. This is verified by the convergence result in Theorem 7.2, which, however, is proven only under the conditions (7.6), implying at least continuity of q and $q f_\alpha$. An inspection of the proof of Theorem 7.2 reveals that these assumptions are needed in order to apply Theorem 7.1 to guarantee that $\|v_\alpha - Q_N v_\alpha\|_0$ exhibits the same asymptotic decay behavior as $\|v_\alpha - P_N v_\alpha\|_0$, where $Q_N : H^\mu_\star(C_{\tilde{r}}) \to T_{\star,N}$ is the interpolation projection with respect to the grid $\mathcal{G}_N = \mathbb{Z}^2_N \odot h_N$ and $P_N : H^\mu_\star(C_{\tilde{r}}) \to T_{\star,N}$ is the orthogonal projection. Details can be found in the proof of Theorem 8.5.3 in [61], for the general idea see also Theorems 8.2.1 and 8.3.1 therein. This relation between the error decays for the orthogonal and the interpolation projection is not guaranteed anymore for discontinuous functions (which still permit a point evaluation such that interpolation remains well-defined). We will show now that this problem can be avoided by computing an approximate solution to (7.3) in a slightly different way. Again, we start from the $\star$-periodic equation (7.5), but consider it only within the medium,

$$u_\alpha(x) = f_\alpha(x) + k_0^2 \int_\Omega K_\alpha(x-y)\, q(y)\, u_\alpha(y)\, \mathrm{d}y, \qquad x \in \Omega. \tag{7.26}$$

We recall that with $f_\alpha = u^i_\alpha$ in Ω the function u_α represents the total field in the direct scattering problem. Before, we multiplied equation (7.5) with the contrast to obtain ($\star$-LSE), following the intention to 'isolate' the field u_α in $\Omega = (\operatorname{supp} q \cap \overline{\Pi})^\circ$, which is the actual relevant region contained in C_r. Opposed to this, in (7.26) we consider the fields by declaration in Ω only. By not multiplying with the contrast q, we prevent a loss of regularity in the case of a discontinuous contrast, comparing u_α with $w_\alpha = q u_\alpha$. Now, we define the bounded operators $D : L^2(\Omega) \to L^2(C_{\tilde{r}})$ and $R : L^2(C_{\tilde{r}}) \to L^2(\Omega)$ by

$$Du = \begin{cases} u & \text{in } \Omega \\ 0 & \text{in } C_{\tilde{r}} \backslash \Omega \end{cases} \qquad \text{and} \qquad Ru = u\big|_\Omega$$

and introduce the new integral operator $\widetilde{\mathcal{K}_\alpha} = R \circ \mathcal{K}_\alpha \circ D$. This operator is bounded as a mapping from $L^2(\Omega)$ to $H^2_\alpha(\Omega)$ and thus compact as a mapping in $L^2(\Omega)$. Moreover, let E_α be a bounded linear extension operator $E_\alpha : H^2_\alpha(\Omega) \to H^2_\star(C_{\tilde{r}})$, which maintains the α-quasi-periodicity. We do not prove rigorously here that such an operator exists, but outline a rough scheme for a proof. Think of Ω as a subset of a torus which represents $C_{\tilde{r}}$, with the vertical parts of the boundary $\partial C_{\tilde{r}}$ glued together and the horizontal parts likewise. Identify the functions in $H^2_{\text{per}}(\Omega)$ with suitable counterparts living on this subset of the torus. For the definition of (and results for) Sobolev spaces on Riemannian manifolds, we refer to [7, Chapter 2]. Under our assumptions on Ω, for the space of these functions an extension operator can be constructed which maps into a Sobolev space of functions on the whole torus, see also Remarks 5.23 in [2] and Appendix A in [53]. To this operator there corresponds an extension operator from $H^2_{\text{per}}(\Omega)$ to the space of H^2-regular functions which are $(2\pi, 2\tilde{r})$-periodic. Combining this with the multiplication operator M_α, $(M_\alpha u)(x) = e^{\mathrm{i}\alpha\cdot x} u(x)$, then yields an operator E_α with the required properties. Let now $w_\alpha \in H^2_\star(C_{\tilde{r}})$ denote a solution to the equation

$$w_\alpha(x) = (E_\alpha f_\alpha)(x) + k_0^2 E_\alpha \left(R \int_\Omega K_\alpha(\cdot - y)\, q(y)\, w_\alpha(y)\, \mathrm{d}y \right)(x), \qquad x \in \overline{C_{\tilde{r}}}. \tag{7.27}$$

It is easy to show that if the assumptions of Theorem 7.2 are fulfilled, then as well as ($\star$-LSE) the equations (7.26) and (7.27) are uniquely solvable and, by the linearity of E_α, the solution w_α to (7.27) equals $E_\alpha u_\alpha$ where u_α solves (7.26). We want to point out that only $w_\alpha|_\Omega = u_\alpha$ has a physical meaning in the context of our scattering problem. The artificial extension provided by E_α serves to obtain a continuous function w_α which encapsulates the physical field in Ω and is accessible to collocation on $C_{\tilde{r}}$. Due to the $\star$-periodicity, it is guaranteed that w_α has a continuous extension to $\partial C_{\tilde{r}}$. This function can be approximated in $T_{\star,N}$ by the solution to the collocation equation

$$w_{\alpha,N} = Q_N E_\alpha f_\alpha + k_0^2 Q_N E_\alpha \widetilde{\mathcal{K}_\alpha}(q R w_{\alpha,N}) \tag{7.28}$$

with respect to the grid $\mathcal{G}_N$, where for the ease of notation q is considered as a function on Ω. The focus here lies on a good approximation of $w_\alpha|_\Omega$ by the restriction to Ω of a function in $T_{\star,N}$. We prove the following convergence result for the solution $w_{\alpha,N}$ to (7.28) in a similar manner as Theorem 7.2.

Theorem 7.3. *Assume that $q \in L^\infty(\Omega)$ and $f_\alpha \in H^2_\alpha(\Omega)$. Let the homogeneous problem corresponding to (7.3) with $f_\alpha \equiv 0$ have only the trivial solution. Then* *equation (7.27) has a unique solution $w_\alpha \in H^2_\star(C_{\tilde{r}})$, the collocation equation (7.28) has a unique solution $w_{\alpha,N} \in T_{\star,N}$ for sufficiently large N, and*

$$\|w_{\alpha,N} - w_\alpha\|_\lambda \le cN^{\lambda-2}\|w_\alpha\|_2, \qquad 0 \le \lambda \le 2,$$

with $c > 0$ independent of α and N. As before, $\|\cdot\|_\lambda$ denotes the norm of $H^\lambda(C_{\tilde{r}})$. In particular, $w_{\alpha,N}|_\Omega$ converges to $w_\alpha|_\Omega = u_\alpha$ in $H^\lambda_\alpha(\Omega)$ with $\lambda < 2$ for $N \to \infty$.

Proof. We recall that the equations (7.3) and (7.5) are equivalent with respect to existence and uniqueness of a solution. Restating the equations (7.5) and (7.27) as

$$u_\alpha(x) = f_\alpha(x) + k_0^2(\mathcal{K}_\alpha \circ D)(qRu_\alpha)(x), \qquad x \in \overline{C_{\tilde{r}}}, \tag{7.29}$$
$$w_\alpha(x) = (E_\alpha f_\alpha)(x) + k_0^2 E_\alpha\big((R \circ \mathcal{K}_\alpha \circ D)(qRw_\alpha)\big)(x), \qquad x \in \overline{C_{\tilde{r}}}, \tag{7.30}$$

respectively, a straightforward argumentation using the linearity of E_α shows that also these equations are equivalent with respect to existence and uniqueness of a solution. In particular, if (7.29) has only the trivial solution for $f_\alpha \equiv 0$, then also (7.30) has only the trivial solution for $f_\alpha \equiv 0$. Now, we define the operator

$$\mathcal{K}^\diamond_\alpha = E_\alpha \circ \big((R \circ \mathcal{K}_\alpha \circ D)q\big) \circ R = E_\alpha \circ \big(\widetilde{\mathcal{K}_\alpha q}\big) \circ R,$$

such that (7.30) reads $w_\alpha = E_\alpha f_\alpha + k_0^2 \mathcal{K}^\diamond_\alpha w_\alpha$ in $\overline{C_{\tilde{r}}}$. It follows directly from the definitions of the operators that $\mathcal{K}^\diamond_\alpha$ is bounded as a mapping from $L^2(C_{\tilde{r}})$ to $H^2_\star(C_{\tilde{r}})$ and hence is compact as an operator in $L^2(C_{\tilde{r}})$. Compare the argumentation for the operator $q\mathcal{K}_\alpha$ in the proof of Theorem 7.2. Again by the Riesz theory, $\mathrm{id} - k_0^2\mathcal{K}^\diamond_\alpha \in \mathcal{L}(L^2(C_{\tilde{r}}))$ is boundedly invertible. Moreover, by Theorem 7.1 with $\mu = 2$ and $\lambda = 0$, there holds $\|\mathcal{K}^\diamond_\alpha - Q_N\mathcal{K}^\diamond_\alpha\|_{\mathcal{L}(L^2(C_{\tilde{r}}))} \to 0$ for $N \to \infty$. Hence, for sufficiently large N there exists $(\mathrm{id} - k_0^2 Q_N \mathcal{K}^\diamond_\alpha)^{-1} \in \mathcal{L}(L^2(C_{\tilde{r}}))$, and its norm is uniformly bounded in N. The rest of the proof goes along the lines of the proof of Theorem 7.2, noting that

$$\begin{aligned}(\mathrm{id} - k_0^2 Q_N \mathcal{K}^\diamond_\alpha)(w_{\alpha,N} - w_\alpha) &= Q_N E_\alpha f_\alpha - w_\alpha + k_0^2 Q_N \mathcal{K}^\diamond_\alpha w_\alpha \\ &= Q_N E_\alpha f_\alpha - w_\alpha + k_0^2 Q_N E_\alpha \widetilde{\mathcal{K}_\alpha}(qRw_\alpha) \\ &= Q_N E_\alpha\big(f_\alpha + k_0^2 \widetilde{\mathcal{K}_\alpha}(qRw_\alpha)\big) - w_\alpha \\ &= Q_N w_\alpha - w_\alpha.\end{aligned}$$

□

We point out that Theorem 7.3 asserts the optimal convergence order for $w_{\alpha,N}$ for all contrasts $q \in L^\infty(\Omega)$, but the price we have to pay for this generalization is that we lose the main advantage of the collocation method (7.10), namely the exact and extremely efficient evaluation of the convolution operator $\mathcal{K}_\alpha$ applied to functions in $T_{\star,N}$. Only for these functions the Fourier coefficients with respect to $\{\varphi_{\alpha,j}\}$, $j \in \mathbb{Z}_N^2$, coincide with those obtained by the discrete Fourier transform (7.13) and the restatement of the operator $\mathcal{K}_\alpha$ in (7.14) is exact. In this case, the Fourier coefficients which belong to $j \in \mathbb{Z}^2 \backslash \mathbb{Z}_N^2$ vanish. In (7.28) with $\widetilde{\mathcal{K}_\alpha} = R \circ \mathcal{K}_\alpha \circ D$, however, the argument of $\mathcal{K}_\alpha$ is only in $L^2(C_{\tilde{r}})$.

Numerical treatment for piecewise constant contrasts

In the remainder of the subsection, we deal with contrasts which are piecewise constant on rectangles. Precisely, we assume that the contrast has the form $q = \sum_{l=1}^L q_l \,\mathrm{id}_{\omega_l}$ where $q_l \in \mathbb{C}\backslash\{0\}$, $\omega_l \subseteq \Omega$ are rectangles, and id_{ω_l} denotes the indicator function of ω_l. For such contrasts, an explicit integration over the singularity of the kernel K_α of the operator $\widetilde{\mathcal{K}_\alpha}$ in (7.28) can be avoided for $x \in \Omega \backslash \bigcup_{l=1}^L \partial\omega_l$ by applying the representation formula (3.20) on each ω_l. First, we rewrite the collocation equation (7.28) as

$$\begin{aligned} w_{\alpha,N} &= Q_N E_\alpha f_\alpha + k_0^2 Q_N E_\alpha \widetilde{\mathcal{K}_\alpha}(qRw_{\alpha,N}) \\ &= Q_N E_\alpha f_\alpha + k_0^2 Q_N E_\alpha \left(\sum_{l=1}^L q_l \widetilde{\mathcal{K}_\alpha} R_l w_{\alpha,N} \right) \\ &= Q_N E_\alpha f_\alpha + k_0^2 Q_N E_\alpha \left(\sum_{l=1}^L q_l R \left(T_N \widehat{\mathcal{K}_{\alpha,N}^{(l)}} \cdot T_N \mathcal{F}_{\alpha,N} \underline{w_{\alpha,N}}_N \right) \right), \end{aligned} \tag{7.31}$$

where $R_l : L^2(C_{\tilde{r}}) \to L^2(\Omega)$ denotes the operator

$$R_l u = \begin{cases} u & \text{in } \omega_l \\ 0 & \text{in } \Omega \backslash \omega_l \end{cases}$$

and $\widehat{\mathcal{K}_{\alpha,N}^{(l)}} : \overline{C_{\tilde{r}}} \to \mathbb{C}^{N \times N}$ is a matrix-valued function defined by

$$\left[\widehat{\mathcal{K}_{\alpha,N}^{(l)}}(x) \right]_j = \int_{\omega_l} G_\alpha(x-y)\, \varphi_{\alpha,j}(y)\, \mathrm{d}y, \qquad x \in \overline{C_{\tilde{r}}},\ j \in \mathbb{Z}_N^2, \tag{7.32}$$

for $l = 1, \dots, L$. This function and the integral operator $\mathcal{K}_\alpha$ from (7.11) are related by

$$\left[\widehat{\mathcal{K}_{\alpha,N}^{(l)}}(x) \right]_j = (\mathcal{K}_\alpha \circ D \circ R_l)(\varphi_{\alpha,j})(x) \qquad \text{for } x \in \overline{C_r} \supset \Omega,\ j \in \mathbb{Z}_N^2.$$

We want to emphasize that we have to use the set $\overline{C_r}$ rather than $\overline{C_{\tilde{r}}}$ here, due to the definition (7.4) of the kernel K_α of $\mathcal{K}_\alpha$. The integral in (7.32) can be interpreted as a x-dependent generalized Fourier coefficient with respect to the function $\varphi_{\alpha,j}$ and the rectangle ω_l. To see this, one should compare (7.32) with the equality

$$\widehat{K_\alpha}(j) = \int_{C_{\tilde{r}}} K_\alpha(y)\,\overline{\varphi_{\alpha,j}(y)}\,\mathrm{d}y = \int_{C_{2r}} G_\alpha(0-y)\,\varphi_{\alpha,j}(y)\,\mathrm{d}y$$

for the Fourier coefficients of K_α, which are used in the discrete form (7.14) of the previously discussed collocation method (7.10). To compute the generalized Fourier coefficients in (7.31) for $x \in \Omega$, we do not apply the convolution theorem to $\mathcal{K}_\alpha$, cp. the relation (7.12). This would involve the Fourier coefficients with respect to $\{\varphi_{\alpha,j}\}$, $j \in \mathbb{Z}^2$, of a discontinuous function which vanishes in $C_{\tilde{r}} \backslash \omega_l$, and these coefficients have a poor decay in $|j|$. Instead, we rely on the representation formula (3.20), which yields

$$\begin{aligned}\left[\widehat{\mathcal{K}^{(l)}_{\alpha,N}(x)}\right]_j = \frac{1}{\lambda_{\alpha,j}}\Big(&-\varphi_{\alpha,j}(x)\,\mathrm{id}_{\omega_l}(x) + \\ &+\int_{\partial\omega_l}\Big(G_\alpha(x-y)\frac{\partial}{\partial\nu}\varphi_{\alpha,j}(y) - \frac{\partial}{\partial\nu_y}G_\alpha(x-y)\,\varphi_{\alpha,j}(y)\Big)\mathrm{d}s(y)\Big)\end{aligned} \qquad (7.33)$$

for $x \in C_{\tilde{r}} \backslash \partial\omega_l$. For a technical reason, we require now, without big loss of generality, that the periodic contrast q can be written as the periodic extension of some $\tilde{q} = \sum_{l=1}^{L} q_l\,\mathrm{id}_{\tilde{\omega}_l}$, $q_l \in \mathbb{C}\backslash\{0\}$, where $\tilde{\omega}_l \subset \mathbb{R}^2$ are rectangles and none of the boundaries $\partial\tilde{\omega}_l$ contains grid points, i.e. $\partial\tilde{\omega}_l \cap \mathcal{G}_N = \emptyset$ for all $l = 1, \ldots, L$. We note that here $\tilde{\omega}_l$ do not need to be subsets of $\Omega = \Omega' \cap \Pi$. In addition, we assume that the x_2-coordinates of the corners of each $\tilde{\omega}_l$ are not multiples of $h_{N,2} = 2c_2/N$. Then the grid $\mathcal{G}_N$ is contained in $\overline{C_{\tilde{r}}} \backslash \bigcup_{l=1}^{L} \partial\tilde{\omega}_l$, and for every grid point $x \in \mathcal{G}_N$ each of the boundaries $\partial\tilde{\omega}_l$ contains at most two points y with $x_2 = y_2$. We illustrate these technical details in Figure 7.1. In this case, we can derive a formal expression for the second term on the right-hand side of (7.31) by means of formula (3.20) and the series representation (7.1) of the Green's function G_α. We do not state this expression here, but only remark that it can be computed to a sufficient accuracy by truncation of the series for G_α at some big index modulus $|z|$, possibly depending on the grid constant $h_N = 2c/N$. At this point, we should also comment on the numerical realization of the extension operator $E_\alpha : H^2_\alpha(\Omega) \to H^2_\star(C_{\tilde{r}})$ (which is not unique). Our current MATLAB implementation for a discretized extension operator follows the scheme

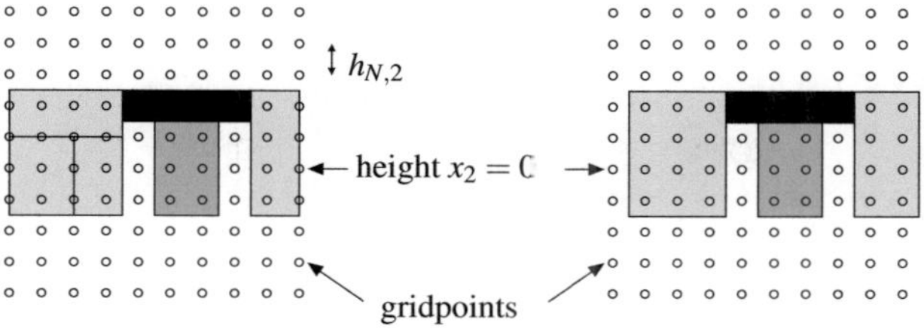

Figure 7.1: *(left)* inadmissible segmentation (with $L = 6$), *(right)* admissible segmentation (with $L = 4$) for the same medium; the shading color indicates the value of the contrast

INPUT: α-quasi-periodic dataset on $\overline{\Omega} \cap \mathcal{G}_N$

(i) Remove the α-quasi-periodicity from the data.

(ii) Enlarge the grid $\mathcal{G}_N$ to $\widetilde{\mathcal{G}}_N = \{j \in \mathbb{Z}^2 : -\frac{N}{2} \leq j_k \leq \frac{N}{2}, k = 1,2\} \odot h_N$ (i.e. insert a column on the left and a row at the bottom of $\mathcal{G}_N$) and frame the data by artificial data points on $F_N = \widetilde{\mathcal{G}}_N \setminus (\overline{\Omega} \cap \mathcal{G}_N)$. Impose $(2\pi, 2\tilde{r})$-periodicity by choosing the data on the first and the last column of the grid $\widetilde{\mathcal{G}}_N$ to be equal and likewise the data on the first and the last row of $\widetilde{\mathcal{G}}_N$.

(iii) Use the MATLAB command '`griddata`' with the option '`linear`' for the discrete linear interpolation of the non-uniformly distributed data on $(\overline{\Omega} \cap \mathcal{G}_N) \cup F_N$. This yields an extension of the data onto the grid $\widetilde{\mathcal{G}}_N$.

(iv) Keep the part on the original grid $\mathcal{G}_N$ and incorporate again the α-quasi-periodicity into the data.

OUTPUT: α-quasi-periodic dataset on $\overline{C_{\tilde{r}}} \cap \mathcal{G}_N = \mathcal{G}_N$, including the given data on $\overline{\Omega} \cap \mathcal{G}_N$

We point out that we have not validated this sort of ad hoc approach in terms of boundedness of the discretized operator and convergence for $N \to \infty$ to a valid operator E_α. Regardless of this evident shortcoming, we observe a clear and appropriate numerical effect.

Instead of with (7.31), we will work with its discrete Fourier representation

$$\widehat{A}_{N^2} T_N \widehat{w_{\alpha,N}} = T_N \widehat{g_{\alpha,N}}, \qquad \widehat{A}_{N^2} = I_{N^2} - k_0^2 \widehat{\mathcal{J}}_{N^2}, \qquad \widehat{g_{\alpha,N}} = \mathcal{F}_{\alpha,N}(\underline{E_\alpha f_\alpha})_N, \tag{7.34}$$

where $\widehat{\mathcal{J}}_{N^2} \in \mathbb{C}^{N^2\times N^2}$ is the matrix representation of the operator $\mathcal{F}_{\alpha,N}(\underline{\mathcal{K}^\diamond_{\alpha N}})_N$: $\mathbb{C}^{N\times N} \to \mathbb{C}^{N\times N}$ and $\mathcal{K}^\diamond_{\alpha N} : \mathbb{C}^{N\times N} \to H^2_\star(C_{\tilde r})$ maps $v \in \mathbb{C}^{N\times N}$ to the function

$$x \longmapsto E_\alpha\left(\sum_{l=1}^{L} q_l R\Big(T_N \widehat{\mathcal{K}^{(l)}_{\alpha,N}}(\cdot)\cdot T_N v\Big)\right)(x), \qquad x \in \overline{C_{\tilde r}}. \tag{7.35}$$

In a similar fashion as for the collocation equation (7.10) for $w_\alpha = q u_\alpha$, one can construct a two-grid iteration scheme for the collocation equation (7.31) for $w_\alpha = E_\alpha u_\alpha$. This can be done equivalently for its Fourier representation (7.34). Doing so, we derive the scheme

$$\begin{aligned}\widehat{w^{(j)}_{\alpha,N}} = \big[I_N + k_0^2 P_{N,M} T_M^{-1} \widehat{A}_{M^2}^{-1} T_M \mathcal{F}_{\alpha,M}(\underline{\mathcal{K}^\diamond_{\alpha N}})_M\big]\cdot&\\ \cdot\Big[k_0^2\big(\mathcal{F}_{\alpha,N}(\underline{\mathcal{K}^\diamond_{\alpha N}})_N - P_{N,M}\mathcal{F}_{\alpha,M}(\underline{\mathcal{K}^\diamond_{\alpha N}})_M\big)\widehat{w^{(j-1)}_{\alpha,N}} + \widehat{g_{\alpha,N}}\Big]&\end{aligned} \tag{7.36}$$

with $j = 1,2,\ldots$, starting from $\widehat{w^{(0)}_{\alpha,N}} = \widehat{w_{\alpha,M}} = P_{N,M} T_M^{-1} \widehat{A}_{M^2}^{-1} T_M \widehat{g_{\alpha,M}}$. We finish this subsection with the remark that the article [68] discusses an alternative Lippmann-Schwinger solver of cubature type, which can handle general piecewise C^2 contrasts and for which a neat convergence result can be proven, based on the theory developed in [67].

7.1.5 Simulation scheme

The complete scheme for the approximate computation of the scattered field u^s_α in $\overline{C_{\tilde r}}$ (more precisely, on the grid $\overline{C_{\tilde r}} \cap \mathcal{G}_N = \mathcal{G}_N$) for a given incident field $f_\alpha = u^i_\alpha$ in Ω now reads as follows.

For a continuous contrast q:

(i) Compute the Fourier coefficients $\widehat{w_{\alpha,N}}(j)$ for $j \in \mathbb{Z}^2_N$ of the approximation to $w_\alpha = q u_\alpha$ by either

  (a) solving the system (7.16) by some direct solver or

  (b) applying the two-grid iteration scheme (7.25) with a coarsening factor $D = N/M$, stopped after a fixed number J of iterations.

(ii) Compute the Fourier coefficients of the approximate $\star$-periodic scattered field by

$$\widehat{u^s_{\alpha,N}}(j) = k_0^2 \tilde{c}^{-1} \widehat{K_\alpha}(j)\, \widehat{w_{\alpha,N}}(j) \qquad \text{for } j \in \mathbb{Z}^2_N$$

with $\widehat{K_\alpha}(j)$ from (7.19).

(iii) Compute the scattered field u_α^s on the grid $\mathcal{G}_N = \mathbb{Z}_N^2 \odot h_N$ approximately by

$$u_\alpha^s(j \odot h_N) \approx u_{\alpha,N}^s(j \odot h_N) = \left(\mathcal{F}_{\alpha,N}^{-1} \widehat{u_{\alpha,N}^s}\right)(j) \qquad \text{for } j \in \mathbb{Z}_N^2. \tag{7.37}$$

For a piecewise constant contrast $q = \sum_{l=1}^{L} q_l \, \mathrm{id}_{\omega_l}$:

(i) Compute the matrix-valued function

$$\widehat{\mathcal{L}_{\alpha,N}} = \sum_{l=1}^{L} q_l \widehat{\mathcal{K}_{\alpha,N}^{(l)}} \qquad \text{on } \mathcal{G}_N, \tag{7.38}$$

with $\widehat{\mathcal{K}_{\alpha,N}^{(l)}}$ as defined in (7.32). For this, use the series representation (7.1) of the Green's function G_α, truncated at the indices with a fixed modulus S. This function $\widehat{\mathcal{L}_{\alpha,N}}$ is closely related to the discretization of the operator $\mathcal{K}_{\alpha N}^{\diamond}$ on $\mathcal{G}_N$, see (7.35).

(ii) Compute the Fourier coefficients $\widehat{w_{\alpha,N}}(j)$ for $j \in \mathbb{Z}_N^2$ of the approximation to $w_\alpha = E_\alpha u_\alpha$ by either

(a) solving the system (7.34) by some direct solver or

(b) applying the two-grid iteration scheme (7.36) with a coarsening factor $D = N/M$, stopped after a fixed number J of iterations.

(iii) Compute the scattered field u_α^s on the grid $\mathcal{G}_N$ approximately by

$$u_\alpha^s(j \odot h_N) \approx u_{\alpha,N}^s(j \odot h_N) = k_0^2 \, T_N \widehat{w_{\alpha,N}} \cdot T_N \widehat{\mathcal{L}_{\alpha,N}}(j \odot h_N) \qquad \text{for } j \in \mathbb{Z}_N^2. \tag{7.39}$$

We make some final remarks about (7.38). The computation of the matrix-valued function $\widehat{\mathcal{L}_{\alpha,N}}$ is a very expensive task in our current implementation. However, the generalized Fourier coefficients $\widehat{\mathcal{K}_{\alpha,N}^{(l)}}$ do not depend on the values q_l of the contrast, $l = 1, \dots, L$, and the computation might be dramatically accelerated by using a more advanced representation of the Green's function G_α to evaluate the right-hand side of (7.33), cf., e.g., [47]. Since the biggest part of the boundary of any of the rectangles which make up Ω normally is well apart from the singularity of the integrand in (7.33), one might also benefit from a simple numerical integration. Moreover, one could compute $\widehat{\mathcal{K}_{\alpha,N}^{(l)}}$ for a big number L of small rectangles in advance and by that make the time-consuming part of the computation fairly independent of the geometry of a given medium. This is meant in the sense that for a medium whose components are unions of any of the small rectangles, the corresponding generalized Fourier coefficients are sums of precalculated ones.

7.1.6 Numerical examples

To get a first idea of the scattered fields, we present some numerical results now. All computations in this and the next section are carried out with a C/MATLAB-package (written by the author) on a PC with an AMD Athlon 64 3800+ @2,4 GHz and 2 GB RAM, using openSUSE 10.3 and MATLAB R2009a. The computation times for the approximate scattered fields are indicated in the subtitles of the plots below. We will point out in particular the dependence of the plots on the discretization constant N in regard of Theorems 7.2 and 7.3. Also, we will illustrate the difference of the plots for a fixed discontinuous contrast which are produced by the two-grid iteration schemes for (7.16) and (7.34), respectively. We recall that the grid $\mathcal{G}_N = \mathbb{Z}_N^2 \odot h_N$ has the convex hull $H_N = [-\pi + h_{N,1}, \pi] \times [-c_2 + h_{N,2}, c_2]$. For the definitions of sets we will still use the familiar notation for the non-discrete setting, but all plots show the area of H_N only. For simplicity, we let $\tilde{r} = 2r$ here, see the simplified expression (7.20) for the coefficients $\widehat{K_\alpha}(j)$, $j \in \mathbb{Z}^2$, in this case. Moreover, to avoid the approximation of the artificial acoustic near field operator M by the physical near field operator $\widetilde{M}$, discussed in Section 5.1, we use directly the artificial incidence which M refers to. This incidence is given as the superposition of fields generated by complex conjugate point sources on Γ_s, cf. (3.25). Clearly, this choice enters the direct problem in the form of f_α. We make some remarks which supplement the explanation at the beginning of Subsection 7.1.4. To obtain a reasonable approximate solution to ($\star$-LSE) on p. 108 by the collocation method (7.14) for a fixed constant N, the Fourier coefficients of the interpolation projection of the unknown function $w_\alpha = q u_\alpha$ on $T_{\star,N}$ should approximate sufficiently well the Fourier coefficients of the orthogonal projection of w_α on $T_{\star,N}$. Obviously, the type of the contrast q is the crucial factor for this. Although the actual smoothness of q is not observable on the grid $\mathcal{G}_N$, big jumps of q on adjacent grid points worsen the described approximation of the Fourier coefficients. This should be regarded in assessing the computations in this section.

Example 1

To start, we consider the simple example of a homogeneous medium given by

$$\Omega = \left(-\tfrac{N_\Omega}{2}, \tfrac{N_\Omega}{2}\right) h_{\Omega,1} \times \left(-\tfrac{N_\Omega}{8} - \tfrac{1}{2}, \tfrac{N_\Omega}{8} + \tfrac{1}{2}\right) h_{\Omega,2}, \qquad q = \begin{cases} 1 & \text{in } \Omega \\ 0 & \text{in } \Omega^{\text{ext}} \end{cases},$$

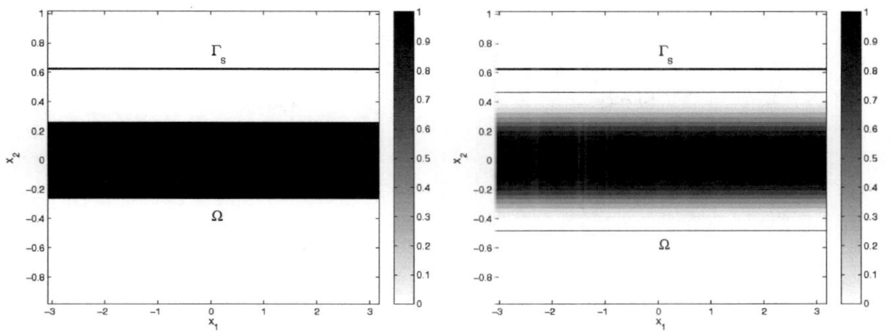

Figure 7.2: *(left)* discontinuous contrast q, *(right)* smoothed contrast $q_{0.2}$

with $N_\Omega = 64$ and $h_\Omega = 2c/N_\Omega$. The rectangle Ω extends over the whole width of the unit cell and represents the characteristic part of a trivially 2π-periodic medium, which is just the strip of height $(N_\Omega/4+1)\,h_{\Omega,2}$. We choose $\tilde{r} = 1 \Rightarrow c = (\pi, 1)$, $\alpha = [0.5, 0]$, and $k_0 = 3$. If not stated otherwise, we let the number $N \leq N_\Omega$ of grid points in each dimension be $N = 64$. Finally, we choose $\Gamma_{s,+} = \{x \in \Pi : x_2 = m_+\}$ for a single-sided incidence and $\Gamma_s = \Gamma_{s,+} \cup \Gamma_{s,-}$ with $\Gamma_{s,\pm} = \{x \in \Pi : x_2 = m_\pm\}$ for a double-sided incidence with $m_\pm = \pm 20\,h_{\Omega,2} = \pm 0.625$. The setting for a single-sided incidence is shown in Figure 7.2 (left). In order to make Theorem 7.2 applicable, we consider also the smoothed contrast given as the convolution of q with the function

$$\varphi_\varepsilon(x) = \begin{cases} \frac{\tilde{c}(\varepsilon)}{\varepsilon^2} \exp\left(-\frac{1}{1-|x|^2/\varepsilon^2}\right) & , |x| < \varepsilon \\ 0 & , |x| \geq \varepsilon \end{cases}.$$

Here, $\tilde{c}(\varepsilon)$ is chosen such that the integral over φ_ε equals one. The parameter ε controls the support of φ_ε and thus the decay of the convolution $q_\varepsilon = q * \varphi_\varepsilon$. Hence, it has direct impact on the approximation quality of the interpolation coefficients, computed by the discrete Fourier transform (7.13). In the choice of ε, we also have to take care that $\operatorname{supp} q_\varepsilon \cap \Pi$ is contained in $C_r = \{x \in \Pi : |x_2| < 1/2\}$, cp. (7.2). For $\varepsilon = 0.2$, q_ε is shown in Figure 7.2 (right), with the straight thin lines indicating its support. We remark that the smoothing affects only the x_2-direction since q continues periodically and has no discontinuity in the x_1-direction. For the plots in Figures 7.3, 7.4, and 7.5 we have chosen a point source incidence $f_\alpha = \overline{G_{-\alpha}(y,\cdot)}$ originating in $y = (0, 20h_{64,2})$ on $\Gamma_{s,+}$. Figure 7.3 (left) shows the real part of the approximate scattered field $u^s_{\alpha,64}$ for the discontinuous con-

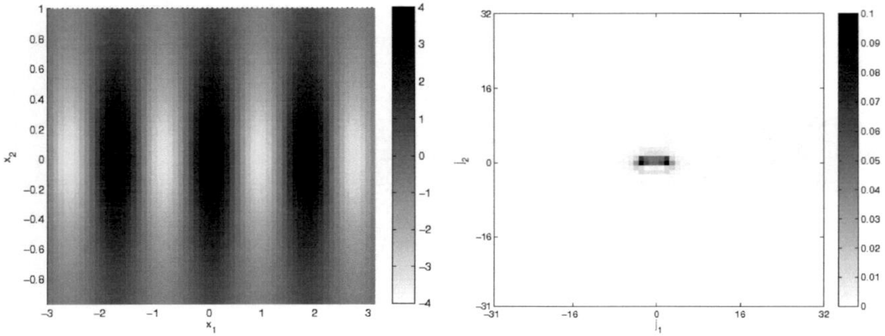

Figure 7.3: *(left)* real part of $u^s_{\alpha,64}$ for q, computed using the iteration scheme (7.36) (1.5 h precomp. + 2 s), *(right)* modulus of $\widehat{g_{\alpha,64}}(j)$ for $j \in \mathbb{Z}^2_{64}$

trast q, computed by means of the two-grid iteration scheme (7.36) with $S = 50$, $D = 2$, and $J = 1$, i.e. stopped after a single iteration. For this example, the precomputation of the function $\widehat{\mathfrak{L}_{\alpha,N}}$ from (7.38) with $S = 50$ takes 1.5 h using our current implementation, whereas the subsequent computation of the scattered field needs 2 s. So, the computation of $\widehat{\mathfrak{L}_{\alpha,N}}$ is the single concern and, up to now, an extreme bottleneck. The starting vector $\widehat{w_{\alpha,32}}$ shows an error of only 2.23 %, which is reduced by the iteration to 0.04 %. The error in the resulting scattered field also amounts to 0.04 %. These data refer to a computation using the exact solution $\widehat{w_{\alpha,64}}$ of the system (7.34) and the 2-norm of the vectors obtained by concatenation of the columns. Crucial for the size of the error in the initial guess is the distribution of the Fourier coefficients $\widehat{g_{\alpha,64}}$ (and the resulting information loss in $\widehat{g_{\alpha,32}}$). For the contrast q, the coefficients $\widehat{g_{\alpha,64}}$ from (7.34) are shown in Figure 7.3 (right). With $M = 32$, the information loss in $\widehat{g_{\alpha,M}}$ amounts to only 1.10 %. Opposed to this, if we apply the scheme (7.25) with $D = 2$ and $J = 1$ to the discontinuous contrast q, the error in the initial guess is 124.54 %, hence the guess is useless. Here, the information about $g_{\alpha_{64}} = \underline{(q f_\alpha)}_{64}$ is spread over many Fourier coefficients, see Figure 7.4 (right), and the coarsening done to compute $\widehat{g_{\alpha,32}}$ causes an information loss of 20.10 %. The single iteration by (7.25) amplifies the error in $\widehat{w^{(j)}_{\alpha,64}}$ to 154.82 %, giving the useless result shown in Figure 7.4 (left), compared to Figure 7.3 (left). This demonstrates the fact that convergence of the two-grid iteration can be established only for sufficiently good initial estimates, inspect the related results in Sections 3.5 and 3.7 in [68]. However, using the exact solution $\widehat{w_{\alpha,64}}$ of the system (7.16) for the discontinuous contrast, we

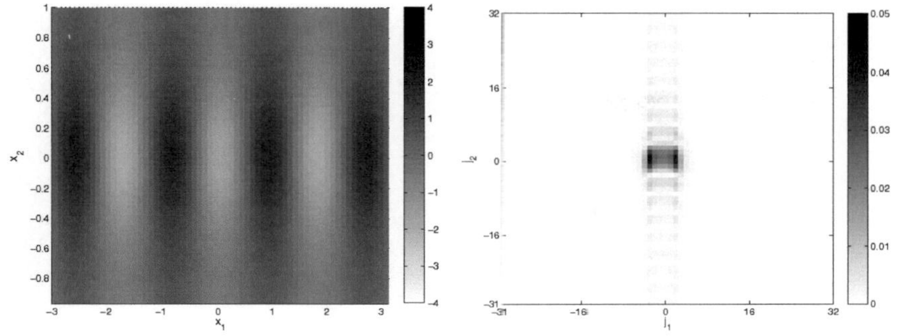

Figure 7.4: *(left)* real part of $u^s_{\alpha,64}$ for q, computed using the iteration scheme (7.25) (3 s), *(right)* modulus of $\widehat{g_{\alpha,64}}(j)$ for $j \in \mathbb{Z}^2_{64}$

obtain a scattered field which differs from the almost exact field, computed by means of the exact solution of the system (7.34), only by 4.77 %. Now, we apply the iteration scheme (7.25) with the same parameters as above to the smoothed contrast $q_{0.2}$. Here, the situation improves a lot, with the relative error in $\widehat{g_{\alpha,32}}$ only 0.22 %. The starting vector $\widehat{w_{\alpha,32}}$ is affected by an error of 0.22 %, and the first iterate is virtually exact with an error below 0.001 %. The same holds true for the resulting scattered field $u^s_{\alpha,64}$, whose real part is plotted in Figure 7.5 (left). These data refer to a computation using the exact solution $\widehat{w_{\alpha,64}}$ of the system (7.16). Our final plots illustrate the evolution of the discrepancy between the Fourier coefficients of the orthogonal projection and the interpolation projection of g_α on $T_{\star,N}$ in dependence of N. The second plot in Figure 7.6 shows the evolution of the error in $w_{\alpha,N}$ and in the approximate scattered field $u^s_{\alpha,N}$ with respect to a reference solution. We consider only powers of two for the value of N and do not take N bigger than 64 here for the following reasons. The evaluation of $u^s_{\alpha,N}$ according to (7.39) involves the function $\widehat{\mathcal{L}_{\alpha N}}$, and the preceding computation of (an approximation to) $\widehat{w_{\alpha,N}}$ via the system (7.34) or the two-grid iteration (7.36) requires the closely related discretization of the operator $\mathcal{K}^\diamond_{\alpha N}$ on the grid $\mathcal{G}_N$. This becomes simply too expensive for $N \geq 128$. To get an idea of the numerical cost, we note that the computation of $\widehat{\mathcal{L}_{\alpha,N}}$ on $\mathcal{G}_N$ comprises $N^2 \cdot N^2 \cdot L$ evaluations of the right-hand side of (7.33), for each of which essentially a (truncated) series needs to be computed. This amounts to about 1.3 billion series computations for this example for $N = 128$. The representation of $\widehat{\mathcal{L}_{\alpha,128}}$ on $\mathcal{G}_N$ as a matrix in $\mathbb{C}^{N^2 \times N^2}$ would occupy about 4 GB of memory for double precision. On

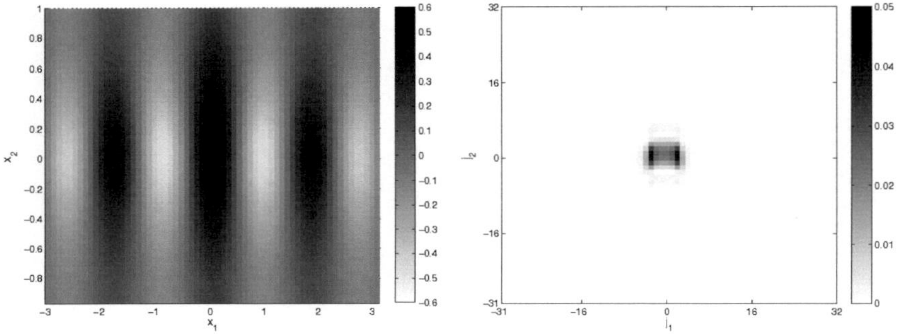

Figure 7.5: *(left)* real part of $u^s_{\alpha,64}$ for $q_{0.2}$, computed using the iteration scheme (7.25) (3 s), *(right)* modulus of $\widehat{g_{\alpha,64}}(j)$ for $j \in \mathbb{Z}^2_{64}$

the other hand, it seems in fact unnecessary to consider a broad range of spectral components, that is a big N, since $w_\alpha = E_\alpha u_\alpha$ is a continuous function on $C_{\tilde{r}}$ and the operator $\mathcal{K}^{\diamond}_{\alpha} = E_\alpha \circ \left(\widetilde{\mathcal{K}_\alpha q}\right) \circ R$ from the initial collocation equation (7.31) is a smoothing operator. Hence, we restrict to $N = 64$. Nevertheless, for the contrast $q_{0.2}$ we can afford to test the result for $N = 128$ of the iteration (7.25) since it does not involve an expensive precomputation like the iteration (7.36). It makes sense in general to consider bigger values of N in (7.16) and (7.25) than in (7.34) and (7.36), since $w_\alpha = q_\varepsilon u_\alpha$ has bigger jumps on adjacent grid points than $w_\alpha = E_\alpha u_\alpha$. However, the change from $w_{\alpha,64}$ to $w_{\alpha,128}$, both evaluated on the grid $\mathcal{G}_{128}$, is minimal for $q_{0.2}$, assuming that the first iterate is almost equal to $\widehat{w_{\alpha,128}}$. Figure 7.7 (left) illustrates the projection errors for $g_\alpha = q_{0.2} f_\alpha$, and Figure 7.7 (right) shows the error in the approximations $w_{\alpha,N}$ with respect to $w_{\alpha,128}$. While the coefficients $\widehat{w_{\alpha,128}}$ are obtained by the two-grid iteration (7.25), the matrices $\widehat{w_{\alpha,N}}$ for the other values of N are exact solutions of the respective system (7.16). Again, also the error in $u^s_{\alpha,N}$ given by (7.37) with respect to $u^s_{\alpha,128}$ is plotted in Figure 7.7 (right). We want to point out that the error decays for $w_{\alpha,N}$ in Figure 7.7 (right) and Figure 7.6 (right) are governed by the error estimates from Theorems 7.2 and 7.3 for $\lambda = 0$, respectively, and provide good numerical evidence for these.

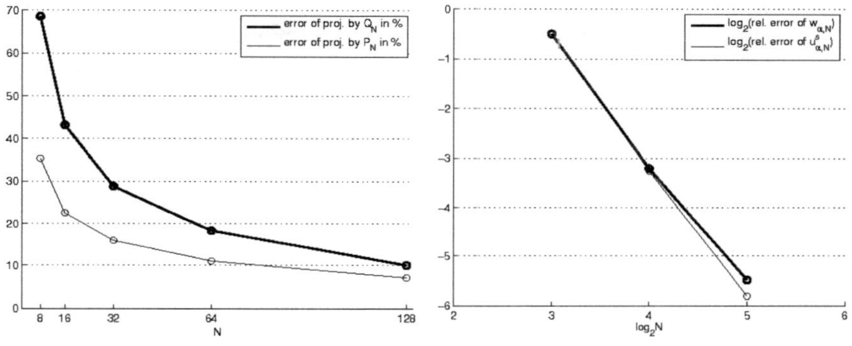

Figure 7.6: *(left)* error in the interpolation projection and the orthogonal projection on $T_{\star,N}$ of $g_\alpha = E_\alpha f_\alpha$, *(right)* error in the approximation $w_{\alpha,N}$ to $w_\alpha = E_\alpha u_\alpha$ with respect to $w_{\alpha,64}$ and error in $u^s_{\alpha,N}$ with respect to $u^s_{\alpha,64}$

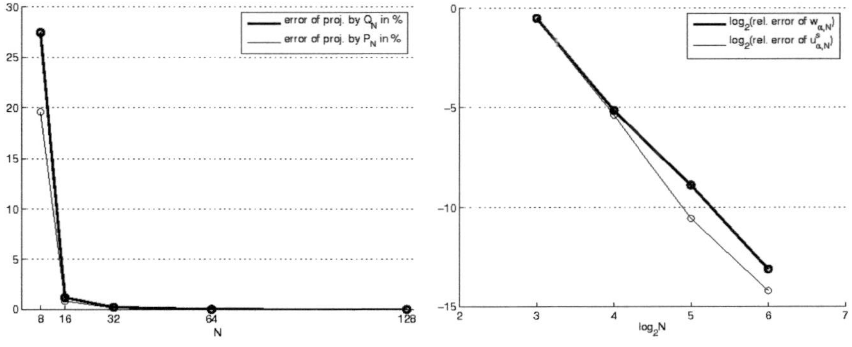

Figure 7.7: *(left)* error in the interpolation projection and the orthogonal projection on $T_{\star,N}$ of $g_\alpha = q_{0.2} f_\alpha$, *(right)* error in the approximation $w_{\alpha,N}$ to $w_\alpha = q_{0.2} u_\alpha$ with respect to $w_{\alpha,128}$ and error in $u^s_{\alpha,N}$ with respect to $u^s_{\alpha,128}$

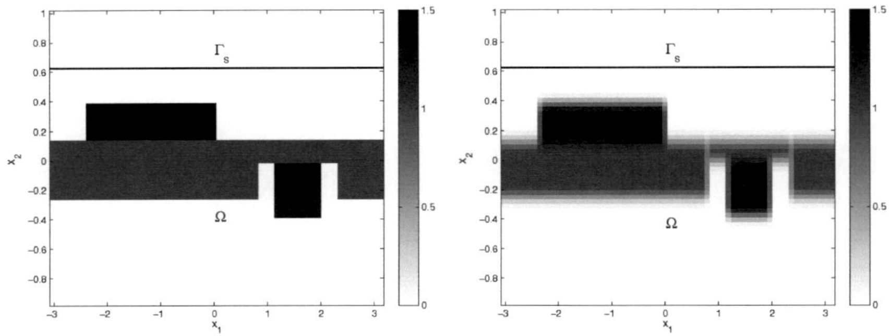

Figure 7.8: *(left)* discontinuous contrast q, *(right)* smoothed contrast $q_{0.1}$

Example 2

To show a more expressive application of our direct solvers, we now consider an example for an inhomogeneous medium with a more interesting geometry than the medium from the first example. Let it consist of the five rectangular components

$$\begin{aligned}
\Omega_1 &= \left(-\tfrac{N_\Omega}{2}, \tfrac{N_\Omega}{8}+\tfrac{1}{2}\right)h_{\Omega,1} \times \left(-\tfrac{N_\Omega}{8}-\tfrac{1}{2}, \tfrac{N_\Omega}{16}+\tfrac{1}{2}\right)h_{\Omega,2}, && q_1 = 1,\\
\Omega_2 &= \left(-\tfrac{3N_\Omega}{8}-\tfrac{1}{2}, \tfrac{1}{2}\right)h_{\Omega,1} \times \left[\tfrac{N_\Omega}{16}+\tfrac{1}{2}, \tfrac{3N_\Omega}{16}+\tfrac{1}{2}\right)h_{\Omega,2}, && q_2 = 1.5,\\
\Omega_3 &= \left[\tfrac{N_\Omega}{8}+\tfrac{1}{2}, \tfrac{3N_\Omega}{8}-\tfrac{1}{2}\right]h_{\Omega,1} \times \left(-\tfrac{1}{2}, \tfrac{N_\Omega}{16}+\tfrac{1}{2}\right)h_{\Omega,2}, && q_3 = \mathrm{i},\\
\Omega_4 &= \left(\tfrac{3N_\Omega}{8}-\tfrac{1}{2}, \tfrac{N_\Omega}{2}\right)h_{\Omega,1} \times \left(-\tfrac{N_\Omega}{8}-\tfrac{1}{2}, \tfrac{N_\Omega}{16}+\tfrac{1}{2}\right)h_{\Omega,2}, && q_4 = 1,\\
\Omega_5 &= \left(\tfrac{3N_\Omega}{16}-\tfrac{1}{2}, \tfrac{5N_\Omega}{16}+\tfrac{1}{2}\right)h_{\Omega,1} \times \left(-\tfrac{3N_\Omega}{16}-\tfrac{1}{2}, -\tfrac{1}{2}\right]h_{\Omega,2}, && q_5 = 1+\mathrm{i},
\end{aligned}$$

where the contrast is given by $q|_{\Omega_l} = q_l$ for $l = 1,\dots,5$ and $q = 0$ outside $\Omega = \bigcup_{l=1}^{5}\Omega_l$. Note that in Ω_3 and Ω_5 energy is absorbed. Again, we choose $N = N_\Omega = 64$, $c = (\pi, 1)$, $\alpha = [0.5, 0]$, $k_0 = 3$, and $\Gamma_{s,\pm}$ as before, with $m_\pm = \pm 0.625$. For the application of the solver based on the collocation system (7.16), we smooth q by means of φ_ε, to meet the regularity assumptions of Theorem 7.2. To ensure that $\operatorname{supp} q_\varepsilon \cap \Pi$ lies in $C_r = \{x \in \Pi : |x_2| < 1/2\}$, we choose $\varepsilon = 0.1$ this time. Figure 7.8 illustrates both settings, showing the modulus of the contrasts q and $q_{0.1}$. We compute the scattered field for the discontinuous contrast by the iteration scheme (7.36) with $D = 2$ and $J = 1$. In the computation of $\widehat{\mathcal{K}^{(l)}_{\alpha,64}}$ for $l = 1,\dots,5$ in (7.38), we choose $S = 50$ for coefficients corresponding to $j \in \mathbb{Z}^2_{64}$ with $\|j\|_\infty > 5$, but $S = 1500$ for $\|j\|_\infty \le 5$. Otherwise, distinctive artifacts occur for this contrast. However, for these values of S and the given medium geometry, the precomputation of $\widehat{\mathcal{L}_{\alpha,N}}$ in (7.38) needs the prohibitive processing time of 10.6 h! Hence,

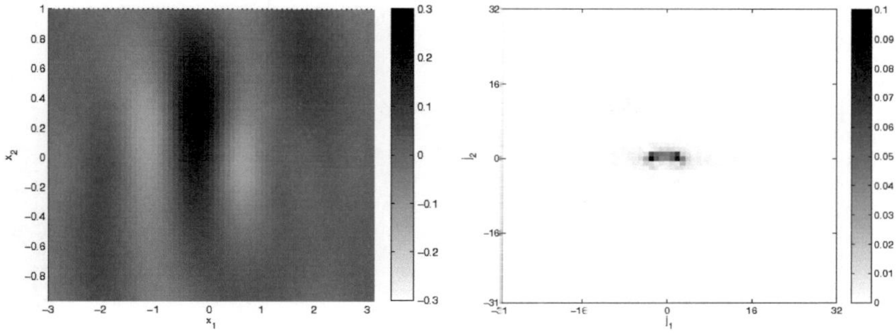

Figure 7.9: *(left)* real part of $u^s_{\alpha,64}$ for q, computed using the iteration scheme (7.36) (10.6 h precomp. + 2 s), *(right)* modulus of $\widehat{g_{\alpha,64}}(j)$ for $j \in \mathbb{Z}^2_{64}$

there is a big need for improvements for this part of the code, like those mentioned at the end of Subsection 7.1.5, in order not to spoil the performance of this solver. The real part of the resulting field is plotted in Figure 7.9 (left), the modulus of the coefficients $\widehat{g_{\alpha,64}}$ in Figure 7.9 (right). For this case, the information loss in $\widehat{g_{\alpha,32}}$ is 2.91 %, and the starting vector $\widehat{w_{\alpha,32}}$ is corrupted by a small error of 3.05 %. This error is diminished by the single iteration to 0.02 %, yielding a relative error in the approximate scattered field of as little as 0.01 %. Analog to the arrangement for the previous example, these data refer to a computation using the exact solution $\widehat{w_{\alpha,64}}$ of the system (7.34) and the 2-norm of the vectors obtained by concatenation of the columns. Opposed to this, if we apply the iteration method (7.25) to the discontinuous contrast q, $\widehat{g_{\alpha,32}}$ carries an error of 25.70 % and $\widehat{w_{\alpha,32}}$ an error of 41.14 % with respect to the exact solution of (7.16). This initial guess seems to lie inside the convergence zone of the iteration (7.25), anyhow, the first iterate deviates from the solution of (7.16) by 14.33 %. The associated scattered field has an error of 12.25 %. Compared to the exact scattered field computed from the solution of the system (7.34), it even has an error of 17.06 %. The results of the application of (7.25) to the smoothed contrast $q_{0.1}$ are shown in Figure 7.11. Here, the loss of information in $\widehat{g_{\alpha,32}}$ is 7.52 % and the initial error is 8.06 %. The iteration reduces the latter to 0.43 %, producing an error in the approximate scattered field of 0.23 %. These figures refer to a computation using the solution $\widehat{w_{\alpha,64}}$ of the system (7.16). The last couple of plots illustrate the discrepancy between the Fourier coefficients for $g_\alpha = q_{0.1} f_\alpha$ and the error decays for $w_{\alpha,N}$ and $u^s_{\alpha,N}$, where the data are computed in the same way as in Example 1. As before, we let the incident field f_α be generated by a complex conjugate point source at $y = (0, 20h_{64,2})$ on $\Gamma_{s,+}$.

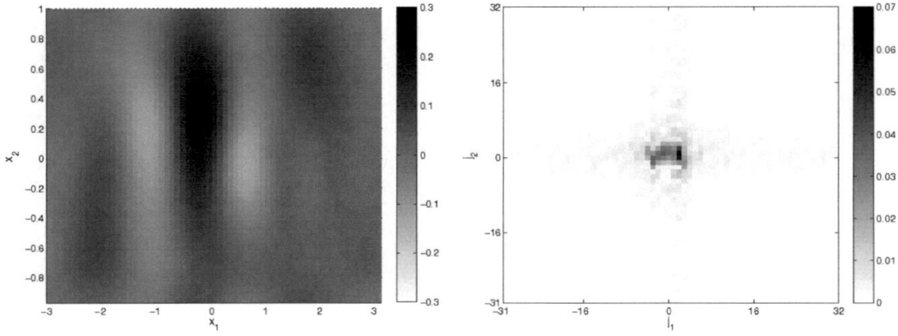

Figure 7.10: *(left)* real part of $u^s_{\alpha,64}$ for q, computed using the iteration scheme (7.25) (3 s), *(right)* modulus of $\widehat{g_{\alpha,64}}(j)$ for $j \in \mathbb{Z}^2_{64}$

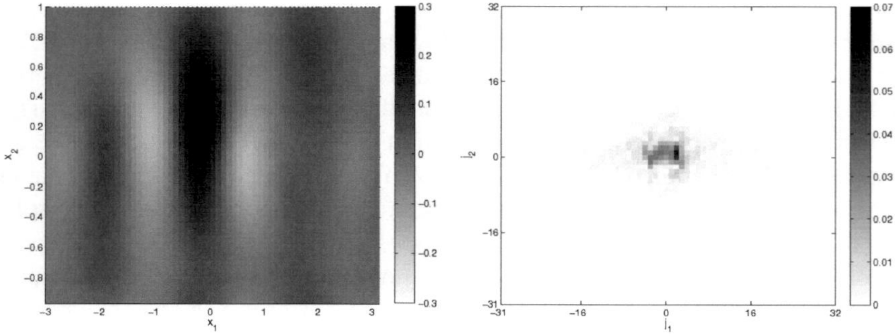

Figure 7.11: *(left)* real part of $u^s_{\alpha,64}$ for $q_{0.1}$, computed using the iteration scheme (7.25) (3 s), *(right)* modulus of $\widehat{g_{\alpha,64}}(j)$ for $j \in \mathbb{Z}^2_{64}$

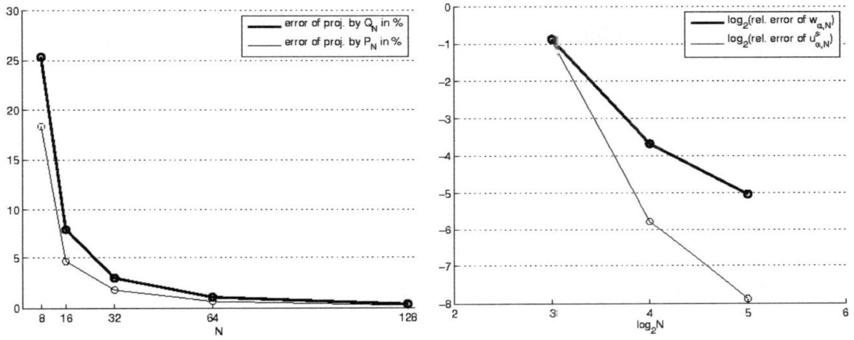

Figure 7.12: *(left)* error in the interpolation projection and the orthogonal projection on $T_{\star,N}$ of $g_\alpha = E_\alpha f_\alpha$, *(right)* error in the approximation $w_{\alpha,N}$ to $w_\alpha = E_\alpha u_\alpha$ with respect to $w_{\alpha,64}$ and error in $u^s_{\alpha,N}$ with respect to $u^s_{\alpha,64}$

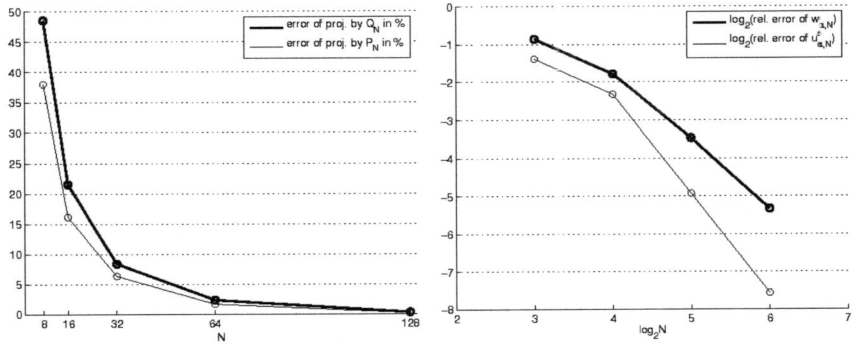

Figure 7.13: *(left)* error in the interpolation projection and the orthogonal projection on $T_{\star,N}$ of $g_\alpha = q_{0.1} f_\alpha$, *(right)* error in the approximation $w_{\alpha,N}$ to $w_\alpha = q_{0.1}\, u_\alpha$ with respect to $w_{\alpha,128}$ and error in $u^s_{\alpha,N}$ with respect to $u^s_{\alpha,128}$

7.2 Inverse problem: Reconstruction of the medium shape

In this final section, we combine the simulation scheme from Subsection 7.1.5 with the results from Sections 6.3 and 6.4 to set up a complete scheme for the reconstruction of the medium shape from simulated acoustic scattering data in 2D. We intend to validate numerically the variant of the Factorization Method which we established in Chapters 3 and 6 as a reconstruction method for periodic, (possibly) inhomogeneous scattering media. After we have formulated the scheme in the next subsection, we do exemplary computations for the media introduced in Subsection 7.1.6 and examine the impact of the parameters on the reconstructions.

7.2.1 Reconstruction scheme (for simulated data)

In real-world experiments, the scattered field is measured on some $\Gamma_s \subset \Omega^{\text{ext}}$, see p. 36 for details on the form of Γ_s. In the numerical simulation, we assume $\Gamma_s \subset \overline{C_{\tilde{r}}}$ and $\Gamma_{s,N} = \Gamma_s \cap \mathcal{G}_N$ to be the non-empty discrete counterpart. These conditions are not essential, but simplify the computations to some extent. The overall reconstruction scheme is as follows:

(i) For every point $y \in \Gamma_{s,N}$, compute the scattered field on the grid $\mathcal{G}_N$ by the procedure described in Subsection 7.1.5 for the incident field generated by a complex conjugate point source at y.

(ii) For each $y \in \Gamma_{s,N}$, extract from the dataset obtained in (i) the values on $\Gamma_{s,N}$ and arrange them in a column vector as the discretization of the response field $u^s_{p,\alpha}(\cdot,y)$ from (3.24) on Γ_s, traversing the grid $\mathcal{G}_N$ from the left to the right and top down. Assemble these column vectors in a matrix $U^s_{p,\alpha}$ in the same order.

(iii) Approximate the integration in (3.24) by the trapezoidal rule. Multiply the sample values in $U^s_{p,\alpha}$ with the corresponding weights. The resulting matrix is the numerical near field operator $M_N : \Gamma_{s,N} \to \Gamma_{s,N}$, our approximation to the artificial acoustic near field operator $M : L^2(\Gamma_s) \to L^2(\Gamma_s)$.

(iv) Compute the Hermitian matrices $\operatorname{Im} M_N$ and $M_{N,\sharp} = |\operatorname{Re}(e^{\mathrm{i}t} M_N)| + \operatorname{Im} M_N$ under the conditions (6.7) and (6.8), respectively.

(v) Compute the eigensystem $\{(\sigma_n, \phi_n)\}_{n\in\mathbb{N}}$ of the square root of the respective matrix from step (iv) and evaluate the Picard series with the discretized probe function $(\psi_{\alpha,z} = G_\alpha(z,\cdot))|_{\mathcal{G}_N}$ for all $z \in \mathcal{G}_N$, cp. (6.11). Truncate the Picard series either at a fixed index T or the index T_σ which belongs to the last eigenvalue σ_n not falling below a certain threshold σ. Plot the reciprocals of the values of the truncated series on $\mathcal{G}_N$ and colorize the convex hull $H_N = [-\pi + h_{N,1}, \pi] \times [-c_2 + h_{N,2}, c_2]$ suitably. The resulting figure gives a rough illustration of the shape of the medium in H_N, according to the criterion from Theorem 6.6 and Theorem 6.10, respectively.

Here, steps (i)–(iii) provide simulated scattering data in the form of the numerical near field operator. Due to the artificial incident field used, this operator can not be composed directly from measurements in practice, but has to be approximated by means of the (approximate) physical near field operator $\widetilde{M}$ according to the convergence relation (5.8). In the computation of the incident fields in step (i) and of the probe functions in step (v), the Green's function G_α is evaluated based on its Ewald's representation. For the derivation and analysis of this representation, we refer to [26, 36] (for the 3D case) and also the forthcoming paper [6] (for the 2D and 3D cases). A survey-like discussion of various techniques to derive a convenient expression for the Green's function (for the 2D case), including new results for Ewald's method, can be found in [51]. We choose this approach here mainly for the purpose of a good accuracy in the neighborhood of a singularity of G_α, but also for efficiency. Since the 'probing grid' for the points z and the 'computation grid' $\mathcal{G}_N \ni x$ are chosen the same in our scheme, we take a very small value for $x - z$ to imitate the singularity of $\psi_{\alpha,z}(x)$ at $x = z$. Finally, we note that in order to identify the medium Ω one might restrict in step (v) to evaluating the truncated Picard series on the subset $\overline{C_r} \cap \mathcal{G}_N$ of the grid, due to the condition $\overline{\Omega} \cap \Pi \subset C_r$, cf. (7.2). We compute the series on the whole grid $\mathcal{G}_N$ only to maintain the plotting area H_N, used in the previous section.

7.2.2 Numerical examples

Let us check the above scheme with the scattering media from Subsection 7.1.6. We are going to examine the sensitivity of the reconstructions to the phase shift α, the wave number k_0, the eigenvalue threshold σ, the measurement height $|m_\pm|$, and to noise in the numerical near field operator M_N. All reconstruction plots below are normalized to fit into the value range $[0, 1]$.

Example 1

We start and apply the scheme for the reconstruction of the discontinuous contrast from Example 1. We use the iterative solver (7.36) with $D = 2$ and $J = 1$ in the

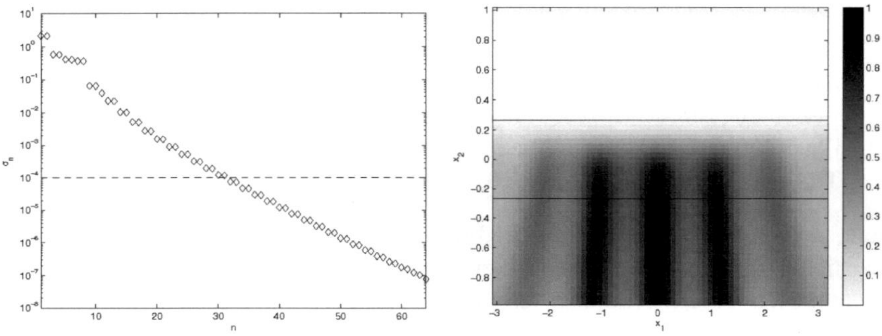

Figure 7.14: single-sided incidence: *(left)* eigenvalues σ_n of $M_{N,\sharp}^{1/2}$, *(right)* reconstruction of the support of q for $T_{\text{1e-4}} = 31$ (18 s)

simulation of the direct problem and the threshold $\sigma = 10^{-4}$ in step (v). Since the contrast is real-valued, we take $t = 0$ in step (iv), the condition (6.8) being satisfied with $c_0 = 1$. For a first numerical experiment, we choose the single-sided incidence from $\Gamma_{s,+} = \{x \in \Pi : x_2 = 0.625\}$; we will comment on this in a moment. Figure 7.14 (left) shows, on a logarithmic scale, the eigenvalues of the numerical operator $M_{N,\sharp}^{1/2}$ for this case, where the dashed line indicates the truncation level σ. The corresponding reconstruction of the contrast is plotted in Figure 7.14 (right). The straight lines mark the boundary of the medium and allow to assess the reconstruction quality. If instead we use the double-sided incidence from $\Gamma_s = \Gamma_{s,+} \cup \Gamma_{s,-}$, we obtain the result shown in Figure 7.15. These reconstructions illustrate the fact that our Factorization Method yields valid results only for a double-sided incidence in general. The reason is that the correspondence argument used in the proof of Theorem 6.1 fails for a single-sided incidence if there is no connecting path between R_+ and R_- in $\Omega^{\text{ext}} = \Pi \backslash \overline{\Omega}$. It is therefore suggested by theory that the bottom part of the shape of Ω is not identified in Figure 7.14 (right). From now on, we work with a double-sided incidence only. The application of the scheme to the smoothed contrast $q_{0.2}$, using the iterative solver (7.25) in step (i) and the threshold $\sigma = 10^{-4}$ in step (v), produces the plots in Figure 7.16. Here, the reconstruction deteriorates near by the boundary of the medium, compare the reconstruction in Figure 7.15 (right). The reason for this numerical effect is that, in the proof of the range criterion, the norm of the preimage g from (6.3) for $z \in \Omega$ rises when z approaches $\partial\Omega \cap \Pi$ since the contrast $q_{0.2}$ decays to zero there. We continue and check the dependence of the reconstruction on α and

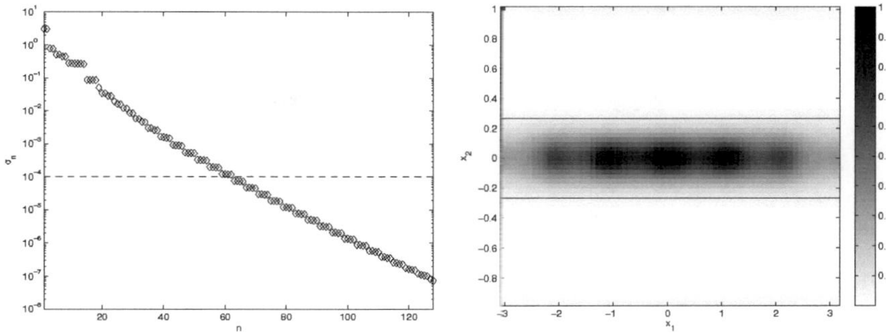

Figure 7.15: double-sided incidence: *(left)* eigenvalues σ_n of $M_{N,\sharp}^{1/2}$, *(right)* reconstruction of the support of q for $T_{1e\text{-}4} = 62$ (35 s)

k_0 for $q_{0.2}$. In each test, we change a single parameter and let the others be fixed to the values chosen above. As indicated by the plot for $\alpha = [0.05, 0]$ in Figure 7.17 (left), the phase shift seems not to have a big impact, which is in accordance with an observation made in [4] (for smaller changes in α than here). A change of the wave number to $k_0 = 1$, however, strongly affects the reconstruction, see Figure 7.17 (right). The associated wavelength $2\pi/k_0 = 2\pi$ is too big to make the medium clearly visible. Similar results for scattering from bounded media are obtained in [49], Section I-6.

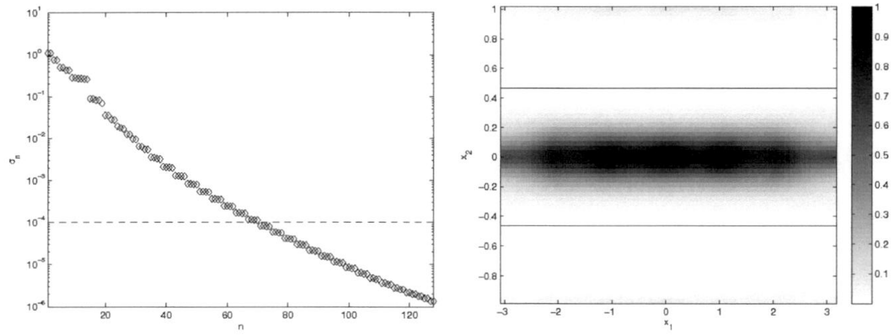

Figure 7.16: double-sided incidence: *(left)* eigenvalues σ_n of $M_{N,\sharp}^{1/2}$, *(right)* reconstruction of the support of $q_{0.2}$ for $T_{1e\text{-}4} = 70$ (35 s)

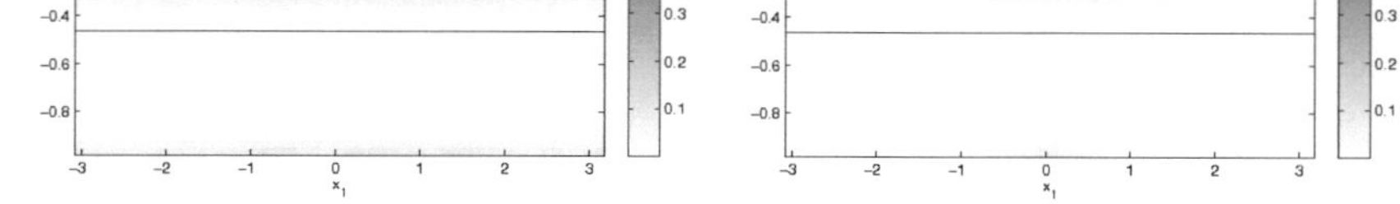

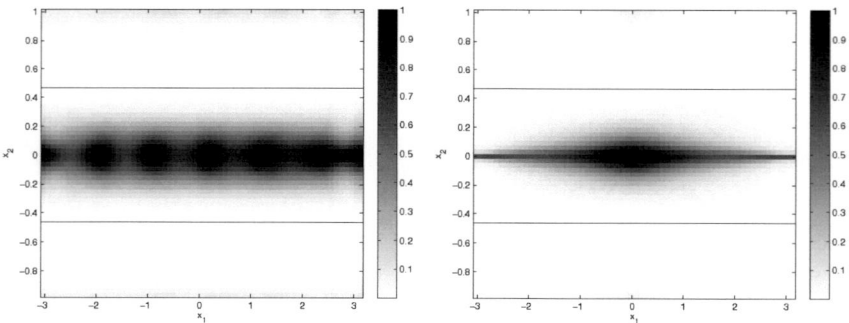

Figure 7.17: double-sided incidence: *(left)* reconstruction of the support of $q_{0.2}$ for $\alpha = [0.05, 0]$ (30 s), *(right)* reconstruction of the support of $q_{0.2}$ for $k_0 = 1$ (35 s)

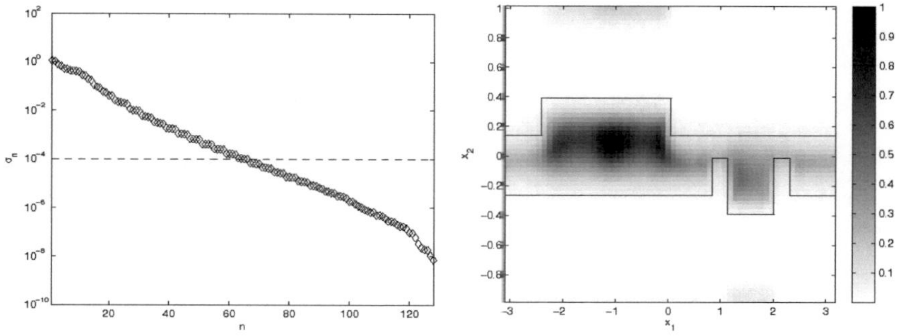

Figure 7.18: double-sided incidence: *(left)* eigenvalues σ_n of $M_{N,\sharp}^{1/2}$, *(right)* reconstruction of the support of q for $T_{1e\text{-}4} = 65$ (35 s)

Example 2

We move on to the inhomogeneous medium from Example 2. As before, we apply the iterative solver (7.36) with $D = 2$ and $J = 1$ in the simulation of the direct problem and the threshold $\sigma = 10^{-4}$ in step (v) of the reconstruction scheme. In step (iv), we let $t = \frac{7}{4}\pi$. We want to point out that this choice is generic in that it lets the condition (6.8) be fulfilled with $c_0 = 1/\sqrt{2}$ whenever $\mathrm{Re}\, q \geq 0$ and $\mathrm{Im}\, q \geq 0$ hold almost everywhere. The result for these parameters is shown in Figure 7.18 (right), corresponding to the $T_{1e\text{-}4} = 65$ biggest eigenvalues of $M_{N,\sharp}^{1/2}$ indicated on the left. The generic value of t does not yield the best visual result for this example, nonetheless the reconstruction is of high quality, with the straight lines marking the contour of the medium. In particular, it resolves nicely the bump on the upper left and the two narrow slots on the bottom right of the medium. The artifacts which appear at the top and the bottom of the plot (and apparently are reflections at $\Gamma_{s,+}$ and $\Gamma_{s,-}$, respectively) do not compromise the reconstruction since we know that the medium lies inbetween $\Gamma_{s,+}$ and $\Gamma_{s,-}$. Now, we perturb the numerical near field operator M_N by 5 % noise with respect to the Frobenius norm, to imitate the typical error source of noisy measurements. The noise matrix which we take generates a relative discrepancy between the projections of the exact and the noisy version of $M_{N,\sharp}^{1/2}$ onto the subspaces corresponding to the 65 biggest eigenvalues, respectively, of 18.50 %. Figure 7.19 shows the resulting plots. Since the reconstruction is still quite decent, we suppose that primary information is carried by less than 65 eigenpairs of the exact operator $M_{N,\sharp}^{1/2}$ and that

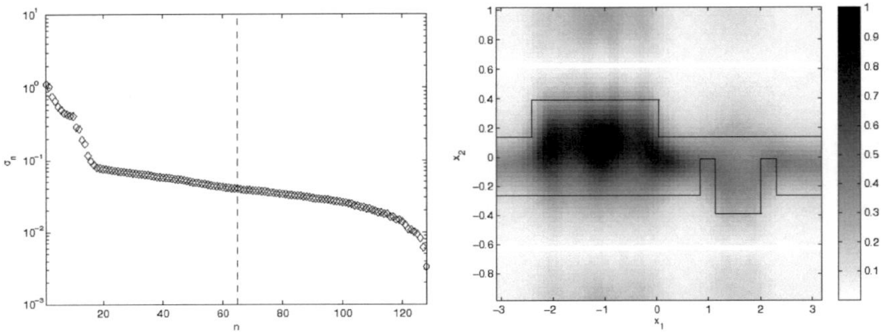

Figure 7.19: double-sided incidence: *(left)* eigenvalues σ_n of noisy $M_{N,\sharp}^{1/2}$, *(right)* reconstruction of the support of q for $T = 65$ (35 s)

the corresponding subset of noisy eigenpairs retains this information sufficiently well. Numerical tests confirm this guess. For our next experiment, we recall that the scattered field u_α^s satisfies a Rayleigh expansion of the form

$$u_\alpha^s(x) = \sum_{z \in Z} u_z^\pm e^{\mathrm{i}(\alpha_z \cdot x \pm \beta_z x_2)} \qquad \text{in } R_\pm, \tag{7.40}$$

where $Z = \mathbb{Z} \times \{0\}$, $\alpha_z = \alpha + z$, $\beta_z = \sqrt{k_0^2 - |\alpha_z|^2} \neq 0$, $R_\pm = \{x \in \Pi : x_2 \gtrless m_\pm\}$, and $u_z^\pm$ are the *Rayleigh coefficients*. It becomes important now to note that the coefficients β_z are real only for finitely many indices $z \in Z$, whereas for all indices z with sufficiently big modulus β_z are purely imaginary with positive imaginary part. Hence, the field u_α^s decomposes in $R_+ \cup R_-$ into finitely many *propagating modes*, the summands in (7.40) which belong to $\beta_z \in \mathbb{R}^+$, and infinitely many *evanescent modes*, which belong to $\beta_z \in \mathrm{i}\,\mathbb{R}^+$ and decay exponentially into R_+ and R_-, respectively. A few wavelengths away from the medium, the information contained in the Rayleigh coefficients of the evanescent modes is hardly observable anymore. Moreover, the higher-frequency components are more likely to be covered by noise. Hence, in the test application of our reconstruction scheme it is pertinent to examine the impact of this information. We change the measurement height now a bit from $m_\pm = \pm 20\,h_{\Omega,2}$ to $m_\pm = \pm 30\,h_{\Omega,2}$ (so that still $\Gamma_s \subset \overline{C_{\tilde{r}}}$, see the comment on p. 134). Using apart from that the same parameters as above, we obtain the plots in Figure 7.20. Obviously, the change affected the reconstruction. We expect the effect to become stronger for bigger $|m_\pm|$. For comparison, we finally consider the smoothed contrast $q_{0.1}$. We apply the iterative solver (7.25)

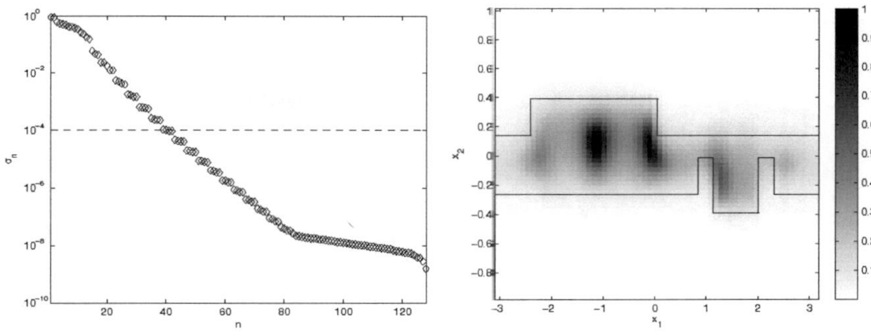

Figure 7.20: double-sided incidence, $m_{\pm} = \pm 30\, h_{\Omega,2}$: *(left)* eigenvalues σ_n of $M_{N,\sharp}^{1/2}$, *(right)* reconstruction of the support of q for $T_{1e\text{-}4} = 40$ (35 s)

with the same configuration as the solver (7.36) before, choosing $m_{\pm} = \pm 20\, h_{\Omega,2}$ first. In Figure 7.21 (right), the straight lines circumscribe the extended support of $q_{0.1}$. (We were a little bit sloppy at the corners.) Again, we notice the decline in the reconstruction quality at the boundary of the medium. It is, however, more constricted than in Figure 7.16 (right), due to the smaller value of ε. Now, let us check what happens when we change $m_{\pm}$. The reconstruction in Figure 7.22 (right) shows a severe deterioration, much stronger than that for the discontinuous contrast. The situation does not improve for a lower truncation level σ. In a following simulation, we removed some of the evanescent modes of the scattered field and found that, for the given set of parameters, the evanescent modes generated by the smoothed contrast carry more information about the shape of the medium than those for the discontinuous contrast. This explains the observed effect. We mention that [4] deals with the scattering of a plane wave from homogeneous periodic media by a boundary integral equation method and illustrates the important role of the evanescent modes in this case.

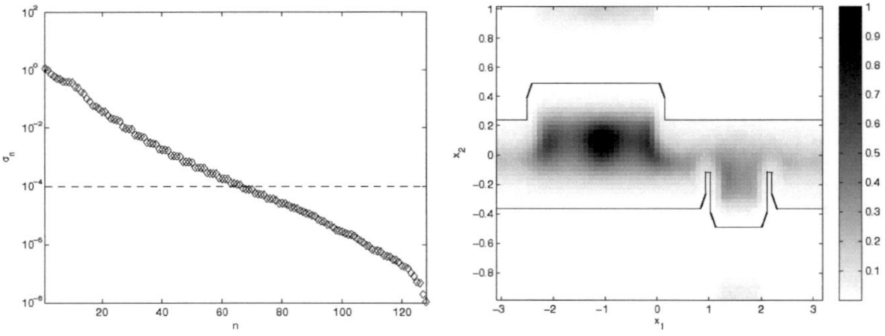

Figure 7.21: double-sided incidence, $m_{\pm} = \pm 20\, h_{\Omega,2}$: *(left)* eigenvalues σ_n of $M_{N,\sharp}^{1/2}$, *(right)* reconstruction of the support of $q_{0.1}$ for $T_{\text{1e-4}} = 66$ (35 s)

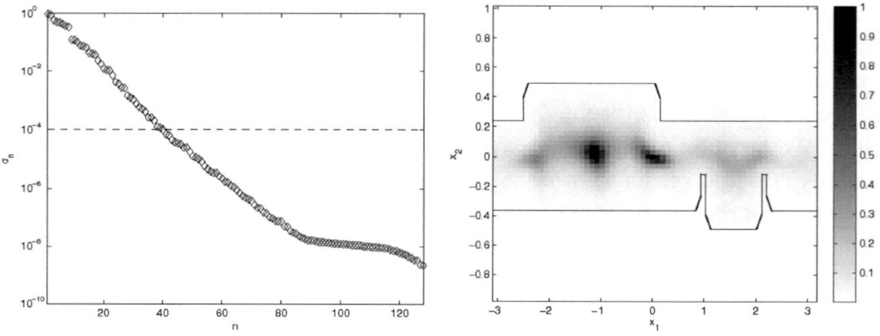

Figure 7.22: double-sided incidence, $m_{\pm} = \pm 30\, h_{\Omega,2}$: *(left)* eigenvalues σ_n of $M_{N,\sharp}^{1/2}$, *(right)* reconstruction of the support of $q_{0.1}$ for $T_{\text{1e-4}} = 40$ (35 s)

7.3 Conclusion

To draw a conclusion, we think that with regard to the ill-posedness of our inverse problem (see Section 6.5) and some small error sources in our computation schemes (see Subsections 7.1.5 and 7.2.1) the reconstruction plots give a good idea of the location and the shape of the scattering medium. The plots might thus be interpreted as numerical evidence for the results from Chapters 3 and 6 which have been used here. The implementation of the Factorization Method is quite simple, and we have shown that it is an efficient and well-founded technique for the shape reconstruction for periodic inhomogeneous media. It can also be used to enhance a full reconstruction method by a quick computation of an initial guess for the contrast. We believe that convincing numerical results can be obtained as well in the electromagnetic case, based on the foundation provided in Chapters 4–6. However, we do not want to conceal that the implementation of the approximation of the physical near field operator (see Chapter 5) requires some additional work. For our direct problem, we developed two variants of a solver for the periodic (acoustic) Lippmann-Schwinger equation. We considered this problem and applied the solvers in order to set up and compute the numerical near field operator. Apart from an expensive precomputation for the second variant, both methods exhibit a very good performance. We hope that, after incorporating the ideas for improvement outlined in Subsection 7.1.5, they can serve as competitive stand-alone solvers for the simulation of time-harmonic scattering from periodic media.

Index

α-quasi-periodicity, 10
analytic Fredholm theory, 33
$\star$-periodicity, 107

boundary value problem
  exterior Dirichlet problem, 44
  exterior electric, 67, 88
  interior electric, 88

conductivity, 1
conical diffraction, 5
contrast, 2

Dirichlet problem, *see* boundary value problem
Dirichlet-to-Neumann operator, 32
dual pairing, 20

evanescent mode, 140

Factorization Method, 39, 93, 101

Green's function
  acoustic case, 33
  electromagnetic case, 56

Helmholtz equation, 5, 29

interior transmission eigenvalue problem
  acoustic, 45
  electromagnetic, 67

Lippmann-Schwinger equation
  $\star$-periodic, 108
  acoustic, 43
  electromagnetic, 65

Maxwell's equations (time-harmonic), 5, 51

near field operator
  acoustic, 36
    numerical, 134
  electromagnetic, 60

periodicity, 2, 10
permeability, 1
permittivity, 1
Picard sequence, 102
potential
  single-layer
    acoustic, 44, 79, 81
  vector
    acoustic single-layer, 90
    electromagnetic double-layer, 66, 85
    electromagnetic single-layer, 89
  volume
    acoustic, 42
    electromagnetic, 63
probe function, 98
propagating mode, 140

quasi-periodic free space Green's function, *see* Green's function
quasi-periodicity, *see* α-quasi-periodicity

radiating

acoustic case, 30
electromagnetic case, 52
radiation condition, *see* Rayleigh expansion
Rayleigh expansion
acoustic case, 29, 140
electromagnetic case, 52
Rayleigh frequency, 5, 30, 52
reciprocity relation, 46
refraction index, 1

single-layer operator, 81
single-layer potential, *see* potential
solution operator
acoustic, 41
electromagnetic, 62

time-harmonic, 2
trace operator
Dirichlet, 21
Neumann, 21
normal, 22
tangential, 22
tangential components, 22
trace space, 19
transmission conditions
acoustic case, 31
electromagnetic case, 53
transverse electric mode, 5
transverse magnetic mode, 5
two-grid iteration scheme, 115, 122

unique continuation principle, 46
unit cell, 3

variational formulation
acoustic case, 31
electromagnetic case, 53
vector potential, *see* potential
volume potential, *see* potential

wave number, 2

Bibliography

[1] private communication with Dipl.-Math. Thomas Gauss, 2008/07/25.

[2] R. A. ADAMS AND J. J. F. FOURNIER, *Sobolev Spaces*, Pure and Applied Mathematics, Elsevier Science (Academic Press), 2nd ed., repr. ed., 2005.

[3] T. ARENS, *Scattering by Biperiodic Layered Media: The Integral Equation Approach*, 2010. Habilitation treatise. Karlsruhe Institute of Technology (KIT).

[4] T. ARENS AND N. GRINBERG, *A complete Factorization Method for scattering by periodic surfaces*, Computing, 75 (2005), pp. 111–132.

[5] T. ARENS AND A. KIRSCH, *The factorization method in inverse scattering from periodic structures*, Inverse Problems, 19 (2003), pp. 1195–1211.

[6] T. ARENS, A. LECHLEITER, K. SANDFORT, AND S. SCHMITT, *Analysing Ewald's Method for the Evaluation of Green's Functions for Periodic Media*. submitted to *IMA Journal of Applied Mathematics*.

[7] T. AUBIN, *Some Nonlinear Problems in Riemannian Geometry*, Springer-Verlag, 1998.

[8] G. BAO, *Variational approximation of Maxwell's equations in biperiodic structures*, SIAM J. Appl. Math., 57(2) (1997), pp. 364–381.

[9] ——, *Inverse and optimal design problems in diffractive optics*, in Recent Development in Theories & Numerics, International Conference on Inverse Problems, 2003, pp. 37–46.

[10] G. BAO, L. COWSAR, AND W. MASTERS, eds., *Mathematical Modeling in Optical Science*, SIAM, Frontiers in Applied Mathematics, 2001.

[11] W. L. BRIGGS AND V. E. HENSON, *The DFT: An Owner's Manual for the Discrete Fourier Transform*, SIAM, 1995.

[12] A. BUFFA AND P. CIARLET JR., *On traces for functional spaces related to Maxwell's Equations. Part I: An integration by parts formula in Lipschitz Polyhedra*, Math. Meth. Appl. Sci., 24(1) (2001), pp. 9–30.

[13] A. BUFFA, M. COSTABEL, AND D. SHEEN, *On traces for $H(\mathrm{curl},\Omega)$ in Lipschitz domains*, J. Math. Anal. Appl., 276(2) (2002), pp. 845–867.

[14] A. BUFFA AND R. HIPTMAIR, *Galerkin Boundary Element Methods for Electromagnetic Scattering*, in Topics in Computational Wave Propagation – Direct and Inverse Problems, vol. 31 of Lecture Notes in Computational Science and Engineering, Springer-Verlag, 2003, pp. 83–124.

[15] F. CAKONI AND D. COLTON, *Qualitative Methods in Inverse Scattering Theory. An Introduction*, Springer-Verlag, 2006.

[16] D. COLTON AND R. KRESS, *Integral Equation Methods in Scattering Theory*, John Wiley & Sons, Inc., 1983.

[17] ——, *Inverse Acoustic and Electromagnetic Scattering Theory*, Springer-Verlag, 1998.

[18] J. W. COOLEY AND J. W. TUKEY, *An algorithm for the machine calculation of complex Fourier series*, Math. Comp., 19 (1965), pp. 297–301.

[19] D. C. DOBSON, *A variational method for electromagnetic diffraction in biperiodic structures*, RAIRO Modél. Math. Anal. Numér., 28 (1994), pp. 419–439.

[20] J. ELSCHNER, R. HINDER, F. PENZEL, AND G. SCHMIDT, *Existence, uniqueness and regularity for solutions of the conical diffraction problem*, Math. Mod. Meth. Appl. Sci., 10(3) (2000), pp. 317–341.

[21] J. ELSCHNER AND G. SCHMIDT, *Diffraction in periodic structures and optimal design of binary gratings. Part I: Direct problems and gradient formulas*, Math. Meth. Appl. Sci., 21(14) (1998), pp. 1297–1342.

[22] J. ELSCHNER, G. SCHMIDT, AND M. YAMAMOTO, *An inverse problem in periodic diffractive optics: global uniqueness with a single wavenumber*, Inverse Problems, 19 (2003), pp. 779–787.

[23] J. ELSCHNER AND M. YAMAMOTO, *An inverse problem in periodic diffractive optics: reconstruction of Lipschitz grating profiles*, Applicable Analysis, 81(6) (2002), pp. 1307–1328.

[24] ——, *Uniqueness results for an inverse periodic transmission problem*, Inverse Problems, 20 (2004), pp. 1841–1852.

[25] H. W. Engl, M. Hanke, and A. Neubauer, *Regularization of Inverse Problems*, Kluwer Academic Publishers, 1996.

[26] P. P. Ewald, *Die Berechnung optischer und elektrostatischer Gitterpotentiale*, Ann. Phys., 64 (1921), pp. 253–287.

[27] D. Gilbarg and N. S. Trudinger, *Elliptic Partial Differential Equations of Second Order*, Springer-Verlag, 2001.

[28] V. Girault and P.-A. Raviart, *Finite Element Methods for Navier-Stokes Equations: Theory and Algorithms*, Springer-Verlag, 1986.

[29] I. C. Gohberg and M. G. Kreĭn, *Introduction to the Theory of Linear Nonselfadjoint Operators in Hilbert Space*, vol. 18 of Translations of Mathematical Monographs, American Mathematical Society, 1969.

[30] F. Hettlich, *Iterative regularization schemes in inverse scattering by periodic structures*, Inverse Problems, 18(3) (2002), pp. 701–714.

[31] F. Hettlich and A. Kirsch, *Schiffer's theorem in inverse scattering theory for periodic structures*, Inverse Problems, 13 (1997), pp. 351–361.

[32] T. Hohage, *Iterative Methods in Inverse Obstacle Scattering: Regularization Theory of Linear and Nonlinear Exponentially Ill-Posed Problems*, PhD thesis, Johannes Kepler Universität Linz, published by Rudolf Trauner Verlag, 1999.

[33] ——, *On the numerical solution of a three-dimensional inverse medium scattering problem*, Inverse Problems, 17(6) (2001), pp. 1743–1763.

[34] ——, *Fast numerical solution of the electromagnetic medium scattering problem and applications to the inverse problem*, J. Comput. Phys., 214(1) (2006), pp. 224–238.

[35] J. D. Jackson, *Classical Electrodynamics (third edition)*, John Wiley & Sons, Inc., 1999.

[36] K. E. Jordan, G. R. Richter, and P. Sheng, *An efficient numerical evaluation of the Green's function for the Helmholtz operator on periodic structures*, J. Comput. Phys., 63 (1986), pp. 222–235.

[37] A. KIRSCH, *Diffraction by periodic structures*, in Proc. Lapland Conference on Inverse Problems, L. Päivärinta and E. Somersalo, eds., Springer-Verlag, 1993, pp. 87–102.

[38] ——, *Uniqueness theorems in inverse scattering theory for periodic structures*, Inverse Problems, 10 (1994), pp. 145–152.

[39] ——, *Characterization of the scattering obstacle by the spectral data of the far field operator*, Inverse Problems, 14 (1998), pp. 1489–1512.

[40] ——, *New characterizations of solutions in inverse scattering theory*, Applicable Analysis, 76 (2000), pp. 319–350.

[41] ——, *The MUSIC-Algorithm and the Factorization Method in inverse scattering theory for inhomogeneous media*, Inverse Problems, 18 (2002), pp. 1025–1040.

[42] ——, *The Factorization Method for a class of inverse elliptic problems*, Math. Nach., 278(3) (2005), pp. 258–277.

[43] ——, *An integral equation approach and the interior transmission problem for Maxwell's equations*, Inverse Problems and Imaging, 1(1) (2007), pp. 159–179.

[44] ——, *An integral equation for Maxwell's equations in a layered medium with an application to the Factorization Method*, J. Integral Equations Appl., 19(3) (2007), pp. 333–358.

[45] A. KIRSCH AND N. GRINBERG, *The Factorization Method for Inverse Problems*, vol. 36 of Numerical Mathematics and Scientific Computation, Oxford University Press, 2008.

[46] R. KRESS, *Linear Integral Equations*, vol. 82 of Applied Mathematical Sciences, Springer-Verlag, 1989.

[47] H. KURKCU AND F. REITICH, *Precise evaluation of the periodized Green's function for the Helmholtz equation at high frequencies*, in Proceedings of the Advanced Computational Methods in Engineering (ACOMEN '08), 2008.

[48] A. LECHLEITER, *A regularization technique for the factorization method*, Inverse Problems, 22 (2006), pp. 1605–1625.

[49] ——, *Factorization Methods for Photonics and Rough Surfaces*, PhD thesis, Universität Karlsruhe (TH), published by Universitätsverlag Karlsruhe, 2008.

[50] I. V. LINDELL, *Methods for Electromagnetic Field Analysis*, Oxford University Press, 1992.

[51] C. M. LINTON, *The Green's function for the two-dimensional Helmholtz equation in periodic domains*, J. Eng. Math., 33 (1998), pp. 377–402.

[52] P. MATHÉ AND S. V. PEREVERZEV, *Moduli of continuity for operator valued functions*, Numer. Funct. Anal. Optim., 23 (2002), pp. 623–631.

[53] W. MCLEAN, *Strongly Elliptic Systems and Boundary Integral Equations*, Cambridge University Press, 2000.

[54] A. MEIER, T. ARENS, S. N. CHANDLER-WILDE, AND A. KIRSCH, *A Nyström Method for a Class of Integral Equations on the Real Line with Applications to Scattering by Diffraction Gratings and Rough Surfaces*, J. Integral Equations Appl., 12(3) (2000), pp. 281–321.

[55] P. MONK, *Finite Element Methods for Maxwell's Equations*, Numerical Mathematics and Scientific Computation, Oxford University Press, 2003.

[56] C. MÜLLER, *Foundations of the Mathematical Theory of Electromagnetic Waves*, Springer-Verlag, 1969.

[57] R. PETIT, ed., *Electromagnetic theory of gratings*, Springer-Verlag, 1980.

[58] A. RATHSFELD, G. SCHMIDT, AND B. H. KLEEMANN, *On a fast integral equation method for diffraction gratings*, Commun. Comput. Phys., 1(6) (2006), pp. 984–1009.

[59] M. RENARDY AND R. C. ROGERS, *An Introduction to Partial Differential Equations*, Springer-Verlag, 1993.

[60] W. RUDIN, *Functional Analysis*, International Series in Pure and Applied Mathematics, McGraw-Hill Inc., 2nd ed., 1991.

[61] J. SARANEN AND G. VAINIKKO, *Periodic Integral and Pseudodifferential Equations with Numerical Approximation*, Springer Monographs in Mathematics, Springer-Verlag, 2002.

[62] S. SAUTER AND C. SCHWAB, *Randelementmethoden: Analyse, Numerik und Implementierung schneller Algorithmen*, B.G. Teubner, 2004.

[63] G. SCHMIDT, *On the diffraction by biperiodic anisotropic structures*, Applicable Analysis, 82(1) (2003), pp. 75–92.

[64] ——, *Electromagnetic scattering by periodic structures*, J. Math. Sci., 124(6) (2004), pp. 5390–5406.

[65] B. STRYCHARZ, *An acoustic scattering problem for periodic, inhomogeneous media*, Math. Meth. Appl. Sci., 21(10) (1998), pp. 969–983.

[66] G. VAINIKKO, *The discrepancy principle for a class of regularization methods*, USSR Comput. Math. Math. Phys., 21 (1982), pp. 1–19.

[67] ——, *Multidimensional Weakly Singular Integral Equations, Lecture Notes in Math. 1549*, Springer-Verlag, 1993.

[68] ——, *Fast solvers of the Lippmann-Schwinger equation*, in Direct and Inverse Problems of Mathematical Physics, Kluwer Academic Publishers, 1999, pp. 423–440.

[69] W. WALTER, *Einführung in die Theorie der Distributionen*, Bibliographisches Institut Wissenschaftsverlag, 1994.

[70] D. WERNER, *Funktionalanalysis*, Springer-Verlag, 2004.

[71] C. H. WILCOX, *Scattering Theory for Diffraction Gratings*, Springer-Verlag, 1983.